Weathering Y2K in Canada

*Alan Bibby &
Akiyah Clements*

Be ready
if the lights
go out in
winter

LONE PINE PUBLISHING   THE UNIVERSITY OF ALBERTA PRESS

Published by
The University of Alberta Press
141 Athabasca Hall
Edmonton, Alberta T6G 2E8
    and
Lone Pine Publishing
206, 10426 - 81 Avenue
Edmonton, Alberta T6E 1X5

Weathering Y2K in Canada is a publication for the book trade from the University of Alberta Press.

Printed in Canada 5 4 3 2 1
ISBN 0–88864–334–9

Canadian Cataloguing in Publication Data

Clements, Akiyah, 1947–
    Weathering Y2K in Canada

    Copublished by: Lone Pine Pub.
    Includes bibliographical references.
    ISBN 0–88864–334–9

    1. Emergency management. 2. Survival skills. 3. Year 2000 date conversion (Computer systems)—
Canada. I. Bibby, Alan, 1942– II. Title.
    QA76.76.S64C53 1999   363.34'97   C99–910331–8

∞ Printed on acid-free paper.
Printed and bound in Canada by Quality Color Press, Edmonton, Alberta.

The University of Alberta Press acknowledges the financial support of the Government of Canada
through the Book Publishing Industry Development Program for its publishing activities. The Press also
gratefully acknowledges the support received for its program from the Canada Council for the Arts.

THE CANADA COUNCIL | LE CONSEIL DES ARTS
FOR THE ARTS | DU CANADA
SINCE 1957 | DEPUIS 1957

Canadä

DEDICATION

This book is dedicated to our children and constant teachers:
Kera, Adon, Ken, Andrew, Robin, and Rachel.

ACKNOWLEDGEMENTS

Thanks to our family and friends for enduring the formation of
this book, with special recognition to Hilary and Katherine for
not doubting our sanity during a time when all conversations lead
to Year 2000. We are indebted to editor Glenn Rollans, whose
contribution to the manuscript far exceeded the normal relation-
ship between authors and publisher. We also want to acknowledge
the support of co-publisher Shane Kennedy, who committed early
to the project; and the staff at the University of Alberta Press,
including Leslie Vermeer, Alan Brownoff, Cathie Crooks, and
Andrew Struthers.

CONTENTS

BACKGROUND

The first rule is to keep an untroubled spirit. The second is to look things in the face and know them for what they are.

MARCUS AURELIUS

The Y2K bug is one of the most pervasive problems ever faced by government and industry. And it will continue to affect all of us beyond year's end.

It is no longer news that there may be widespread disruptions caused by the misinterpretation of the date '2000' by computer programs and various electronic devices

Canadians in particular face an added challenge: winter. If the 'Year 2000 bug' disrupts electrical power, heating fuel, communications, food, or water supplies, Canadians will have to cope in the depths of winter until services are returned to normal.

It is difficult to predict what the situation will be like in the early days of the year 2000. Remedial work is progressing to fix the known problems, and certainly the Canadian situation looks better than it did in late 1998. All levels of government in Canada, along with most major businesses and utilities, are attempting to fix the known problems and putting in place contingency plans to deal with the unexpected.

We recommend that you should do the same.

We think that possible Year 2000 disruptions require planning for the kind of real-world situation that Canadians encountered in the 1998 ice storm. Nothing extreme or alarmist, just reasonable preparations for real risks that we face because of our cold climate—not just this winter, but every winter.

Take some simple steps to protect your family, your home, and yourself. *Weathering Y2K in Canada* can help you to meet the New Year with all the optimism and joy that it deserves.

HOW TO USE THIS BOOK

There is a great deal of information out there on the Y2K problem: racks of technical books, scary survivalist fantasies, and literally thousands of websites. Y2K may have generated more bits and bytes than anything else in the information age. But only some of this overwhelming mass of sometimes-contradictory advice can be considered relevant to Canadians.

We have organized this short, practical guide so that you can quickly read what you need and create a work plan for your own situation.

We've tried to make sure that everything we suggest adds to your safety and comfort, and the safety and comfort of others around you. Where you suspect otherwise, you should always err on the side of caution. Don't plan to be warm at the risk of burning down your house, and don't plan to be safe at the risk of harming others.

Read each chapter, with its descriptions of options and priorities. Then, while the information is still fresh in your mind, go to the checklists at the back of the guide and mark each item you've decided to act on.

By the time you have worked through all the chapters, you'll have a complete list of all the tasks that you've decided to accomplish.

WHY PREPARE?

Some Canadian municipalities have now announced that they will base their contingency plans for Y2K on three days without power. How would you cope in your household if there were no electricity for three days?

The outside temperature is -25°C. Your home is cooling by about one degree per hour.

The power failure hit the water company too, so you can't get a drink of water and you can't flush your toilet. Maybe the phones are dead, and banks and stores are either closed or cash-only. You might be in a dark apartment without any way of cooking. And especially if you have a disability, you're feeling vulnerable.

You may have money, so a hotel sounds good, or maybe a plane to somewhere warm. But even if the hotel has power and the planes are flying, lots of other people have had the same idea. And besides, what about your house, your pipes, pets, and plants?

Most Canadians would be in serious trouble very quickly with no heat, no light, and inadequate water and food supplies. Living in a cold climate can turn an inconvenient situation into a life-threatening one.

Take the ice storm that hit Quebec, Ontario, and some northern U.S. states in 1998. At the storm's peak, 1.4 million people had no power; many communities struggled in survival conditions in what became known as the Dark Triangle. Hundreds of thousands of people huddled in homes without heat and some succumbed to hypothermia.

At least 30 deaths were attributed to the ice storm. Even with massive assistance from the rest of the country and emergency teams from the U.S., it took more than a month to restore power to some parts of the storm area. Many people found that they were quite unprepared for the effects of a prolonged storm coupled with the loss of essential utilities.

It seems a fair assumption that most of those who experienced the ice storm will not be caught unprepared again.

Whether you're preparing for winter weather or the unpredictable effects of the Y2K bug, we believe you would rather be reasonably well prepared and relieved if nothing happens than under-prepared and worried that something will.

THE Y2K BUG

Way back in the days when computers were the size of living rooms and programmers used punch-cards to enter code and information, using four digits to write the year seemed a monumental waste of space.

So everyone agreed that '74' could be universally understood to mean '1974.' This is a little like making a calculator that only remembers the last two digits of numbers. As an example, if you take out a three-year mortgage in '96, the bank will automatically remind you of the renewal date in '99. But what happens if you take out a three-year mortgage in '98? The two-digit convention is prone to errors!

The message was out there many years ago that this two-digit representation of the year had serious disadvantages. Unfortunately, many programmers chose to ignore and perpetuate the problem. They wrote two-digit dates into millions—even billions—of lines of code that control financial systems, defence systems, personal computers, and other electronic systems. And then they reused those faulty lines of code to form the foundations of new programs, burying the flaws like needles in haystacks.

Similar problems were built right into computer chips (embedded chips) and scattered to the winds. These faulty 'brains' control some functions of computers, trains, planes, and automobiles, fax machines, VCRs, coffee machines, and other 'smart' devices.

At the stroke of midnight, 31 December 1999, things may start to go awry.

Of course the good news is that many programs and chips have now been fixed, and many won't need to be. The vast majority of computer chips and embedded processors will continue to function perfectly, well beyond the year 2000.

But we just don't know which ones will go wrong. If it's the non-compliant chips in your VCR, big deal. If it's the embedded chip in a device that controls the flow of drugs to a loved one, or the mission-critical chips in a city water purification plant then the situation becomes much more serious.

When devices and programs fail completely, at least you'll know that they've gone wrong. Unfortunately, some may malfunction or produce data errors that won't be discovered for some time, perhaps after being passed to another application or device. If those errors appear in your bank statement or credit-card bill, you will need your own paper records to prove the discrepancy.

Thousands—probably millions—of people around the world are doing their very best to find and fix the problems, or, failing that, to work around them.

At the same time, you can take the simple steps we describe in this book to protect yourself from the unexpected.

> The bright side to Y2K is that we will have 8,000 years to prepare for 5-digits.

RONALD C. STEVENSON,
quoted in the *Globe and Mail* 13 February 1999

THE GLOBAL VIEW

> The apocalyptic crowd sometimes uses an acronym to summarize the forecast of the blackout to eclipse all blackouts as 'TEOTWAWKI—the end of the world as we know it.' On the other hand, there are skeptics like Jim Wilson, the science editor of Popular Mechanics, who 'dismissed Y2K as an urban legend, apparently on the grounds that the computer industry couldn't possibly be that stupid'.

WENDY GROSSMAN,
'Cyberview,' *Scientific American*, October 1998

A lot depends on the U.S.—the engine of the global economy. The official line from the White House is that a great deal needs to be done before the end of 1999.

> We need to make sure that this Y2K bug will be remembered as the last headache of the 20th century, not the first crisis of the 21st.

PRESIDENT BILL CLINTON,
20 January 1999, quoted on the Federal Emergency Management Agency website

In early March 1999, the U.S. Senate Special Committee on the Year 2000 Technology Problem issued a report observing:

> The good news is that talk of the death of civilization, to borrow from Mark Twain, has been greatly exaggerated. The bad news is that Committee research has concluded that the Y2K is very real and that Y2K risk management efforts must be increased to avert serious disruptions.
> The Special Committee conducted extensive research and held numerous hearings in 1998, but still cannot conclusively determine how extensive the

Y2K disruptions will be. The Committee has no data to suggest the United States will experience nation-wide social or economic collapse, but the Committee believes that in some cases Y2K disruptions may be significant. The international situation may be even more tumultuous.

Investigating the Impact of the Year 2000 Problem, 2 March 1999

Canada, along with the rest of the world, will watch closely to see if the U.S. can get its house in order.

Will everything be fixed in time? Definitely not. Will all the 'mission critical' systems be finished on time? Highly unlikely. Will things be missed? Certainly. The task of finding and dealing with all the faulty embedded chips has been likened to trying to change all the light bulbs in Las Vegas in a single afternoon: it's a simple job, but vast and prone to errors.

In January 1999, the *Scientific American* published a table developed by Artemis Management Systems that indicates the 'Bottom Line for Y2K in the U.S.'

	Best Case	Expected Case	Worst Case
Software applications with Y2K problems	10 million	12 million	15 million
Percent of Y2K problems that will not be fixed in time	5%	15%	25%
Infrastructure failures because of Y2K* Power systems	5%	15%	75%
Transportation systems	5%	12%	50%
Telephone systems	5%	15%	65%

* Percent of households that will be affected

In the first quarter of 1999, as systems tests proceed, the predicted 'Best Case' of a five per cent failure rate is proving to be close to the mark.

In the global scheme of things, the level of preparedness in the U.S., Canada, Britain, and Australia is consistently rated in the top tier when compared to the rest of the world.

> *The majority of the planet is lagging well behind and in some countries, the after-effects of the bug could be catastrophic. Several U.S. trading partners are severely behind in their remediation efforts. For example, the GartnerGroup estimates that Venezuela, which is the largest supplier of oil to the U.S., is nine to fifteen months behind the U.S. in its Y2K preparation.*

Report released 2 March 1999 by the U.S. Senate

Especially poorly prepared regions, according to the same report, include Russia and many countries in Africa and South America. Japan's slow progress on Y2K preparations is a special worry for its trading partners.

Canada can't isolate itself from across-border problems or overseas failures. Our trade-dependent economy, like our electrical system, may react to problems outside of our control. The true challenge of the Y2K bug may be the stressed supply chain for food and goods that stretches back to Chile, Japan, Poland, China, and the rest of the world. Other national economies face the same challenge in the coming year, and some economists are predicting a major downturn next year.

HOW BAD WILL IT BE IN CANADA?

> *Will the lights go out? The answer is that no one knows for certain yet what the effects of Y2K will be. The risks that Y2K may impact electrical system operations are real—much like the risks that earthquakes or severe weather could cause electrical outages even before the millennium arrives.*

North American Electric Reliability Council Report, September 1998

Electricity is a lifeblood of our society, so we need to pay special attention to Y2K problems in our electrical systems. Canadian electrical companies express cautious optimism about their state of readiness for the Y2K bug and have conducted some limited tests without problems. But they're also planning for the possibility of darker scenarios because of factors beyond their control.

Electrical generating plants in Canada, the U.S., and Mexico are linked by high-tension lines, and they operate largely as one interconnected system, organized around regional 'interconnection grids.'

> [T]he interconnectedness makes the grid fragile and susceptible to Y2K disruptions. An outage in one part of the grid can cascade causing ripple effects on other parts of the grid. For example, a power generation plant could go out in Maine, affecting power in Florida.
>
> While complete power grid failure and prolonged blackout is highly unlikely, failure of at least some parts of the electric power industry, e.g., local and regional outages, is possible. The 3200 electric facilities are at various stages of remediation.
>
> Report on Utilities released 2 March 1999 by the U.S. Senate

Federal, provincial, and municipal agencies in Canada have drawn up contingency plans to deal with the possibility of widespread power failures.

For example, the Canadian military has set up 'Operation Abacus,' which could be the largest peacetime deployment of armed forces personnel in the country's history. 'Abacus' refers to the traditional Chinese calculating tool that doesn't need power and is foolproof in the right hands.

Over 14,000 mobile-forces personnel, together with about 4,000 reservists, will be on alert for anything that occurs at midnight on 31 December 1999. The navy will have frigates standing by to provide power and communication links.

The Armed Forces released a memo on 15 January 1999 that says, among other things: 'The entire Canadian forces must be prepared for a major operation over a protracted period of time.' The plan, like this book, deals with reasonable contingencies: it is not built around 'a total collapse of Canada's water, hydro, sewer and telecommunications but rather local disruptions.'

To put this in perspective, during the 1998 ice storm, more than 16,000 troops were deployed to assist local resources. In the case of serious, widespread problems across the country, the Armed Forces won't attempt to manage the situation alone. They are relying on provincial and municipal authorities to make their own preparations and handle their own local problems.

In case things do get somehow out of hand, Ottawa has dusted off the Emergencies Act, a successor to the War Measures Act. This is being characterized as prudent planning and doesn't mean that the Act will automatically come into force.

So, how bad will it be in Canada? Really, no one knows. We may experience very few effects from the Y2K bug, or we may have some very serious challenges to contend with.

Those in positions of authority who have responsibility for public safety are developing plans to deal with the worst-case scenario.

As stated earlier, you need to make your own choices, and make your own personal preparations, in a way that's reasonable and achievable for you.

❋ **Further reading**

❋ One of the best all-round articles on Y2K is in the January 1999 issue of *Scientific American*, available at libraries or online at **http://www.sciam.com/1999/0199issue/0199dejager.html**

❋ For good background on Y2K, we recommend a publication from the *Utne Reader:* the *Y2K Citizen's Action Guide*. This booklet sells in some Canadian bookstores for $5.95, or it can be downloaded free at **www.utne.com/y2k** on the web.

❋ *Time* magazine maintains a Y2K site at **http://cgi.pathfinder.com/time/reports/millennium/index.html**

❋ The U.S. President's Council on Year 2000 conversion can be found at **http://www.y2k.gov/**

❋ The Canadian Federal Government Year 2000 Information site, with extensive links, is at **http://www.info2000.gc.ca/**

❋ Industry Canada's 'Strategis' site appears at **http://strategis.ic.gc.ca/**

❋ The full text of the NERC report can be found at **http://www.nerc.com/~y2k/y2k.html**

❋ The Ottawa-based Global Millennium Foundation reports on national and international progress. The former director of

the Year 2000 Program of the Canadian Imperial Bank of Commerce heads it up. The site is at
http://www.globalmf.org/

❋ The full text of the U.S. Senate Report issued on 2 March 1999 is available at **http://www.senate.gov/~y2k/**

FIRST STEPS

The bottom line: Individuals should prepare for limited duration localized failures of services and infrastructure rather than an apocalypse. The type and number of failures will vary geographically and cannot really be predicted.

Year 2000 Risk Assessment and Planning for Individuals,
28 October 1998 by the GartnerGroup,
an international authority on Information Technology
that makes frequent appearances at U.S. Senate Hearings.

WHAT CAN I DO?

Weathering Y2K in Canada offers a 'stay at home' plan, not a 'run for the hills' plan. We recommend that everyone should do what they reasonably can, whether that be preparing for full self-sufficiency for an extended period of time, or simply planning where to go for shelter if necessary, because that's all you can afford.

Preparations will differ according to your personal circumstances:

* if you are in the city or a rural area
* if you own a house
* if you rent an apartment
* if you have surplus income or live from month to month
* if you have special needs.

The more you prepare, the less your world will be disrupted by anything that the Year 2000 can throw at you—up to a point. Make reasonable preparations, but don't turn your world upside down, spending money you can't afford, buying things you'll never have a use for, or taking steps that endanger you rather than adding to your safety and comfort. That is what this guide is all about.

CAN I WAIT TO SEE IF THERE'S A PROBLEM?

No! The sooner you get started, the more likely you are to be able to do everything that you want to do. As the year progresses, the flood of conflicting messages about Y2K will escalate and unprepared people may begin to get anxious. Some of the contingency items that you may want to purchase simply won't be available in the last few months of 1999.

Think through your situation immediately, and begin preparing as soon as you can, and you'll save yourself worry, money, and possibly discomfort or danger.

HOW LONG DO I NEED TO PLAN FOR?

As we mentioned earlier, several Canadian municipalities are basing their plans on a scenario involving 72 hours without power, but there are so many variables in the continental and global scenario that we believe it prudent to plan for longer if you have the resources to do so.

As a starting point, therefore, we recommend that you make plans for:

* **1 week** of warmth and light
* **1 week** of water for drinking and sanitation
* **1 month** of food
* **2 weeks** of cash.

Treat these as guidelines, and vary them according to your own research and your own circumstances. You may decide to extend these times if, for example, you live in an isolated area or somewhere with an extreme climate, or simply because you have the money and the inclination.

Here is some background on the contingency plans that are being developed in Canada:

> *The RCMP has cancelled all leave from 27 December 1999 until 15 March 2000 based on a decision to 'expect the unexpected and work on the worst case scenario.'*
>
> KATHRYN MAY,
> 'RCMP on alert for Y2K Disaster,' *Ottawa Citizen*, 3 October 1998

RCMP Commissioner Phil Murray sent out a memo that drew a parallel to the 1998 ice storm:

> *It is not expected that we will experience "Ice Storm" conditions during this entire period. However, problems probably of a limited duration may occur.*

KATHRYN MAY,
'RCMP on alert for Y2K Disaster,' *Ottawa Citizen*, 3 October 1998

Dave Morreau, who is in charge of the RCMP's Year 2000 project, says that 'cautious Year 2000-ists say it would be prudent to have a couple of weeks' supply' of food and cash on hand.

A '72-hour scenario' appears in many briefings and planning documents. We return to our recommendation to use this as a starting point, and we believe that it is prudent to consider solutions that extend well beyond this time frame.

In the chapters that follow, you will find suggestions that range from short-term fixes to permanent backup systems. For example:

* ❋ you can put together a no-cost plan to create a warm space in your home or spend $1,800 to install a wood-stove
* ❋ you can top up your cupboard with an extra three days' supply of food or put a four-week supply in your pantry.

In summary: we present practical options to consider for weathering a limited period of Y2K disruptions, and we invite you to consider preparations for a longer duration.

WHO ARE YOU PLANNING FOR?

You must decide whether you're planning for just yourself and your immediate family, for your extended family, or even for friends and neighbours if they're in trouble.

You should also consider whether you're planning for adult able-bodied people, or for people with special needs, such as babies, the elderly, or people with disabilities.

Review whether anyone you're planning for has any special medical or dietary requirements.

Finally, consider whether you are planning for pets and other domestic animals.

These considerations will help shape your plans as you work through later sections.

WHERE ARE YOU GOING TO BE?

Decide whether you are going to be at home, at a local party, at festivities in your community, on a holiday, or even in another country when the calendar changes.

You may already have big plans in place for a Year 2000 celebration and be unwilling to change them, or you may be willing to change your plans as part of your overall Y2K strategy.

We recommend that you consider being somewhere where there will be adequate contingency plans in place for Y2K problems, and that, wherever you are, you make plans to ensure the safety and security of your home.

DEALING WITH STRESS

While we can't offer anything that compares to professional help in this area, we do want to reassure you about the anxiety that often develops when preparing for the consequences of the Y2K bug.

The media are carrying a steadily increasing stream of stories about the Y2K bug, and the volume will increase during 1999. Unfortunately, bad news means good business, so the majority of the stories may be biased towards problems rather than solutions.

It's very easy to slip into panic mode, especially since this very specific problem coincides with the end of the millennium, a date with mystical, worrisome connotations for some.

We believe there's a need for reasonable preparations, and the earlier the better. But remember that most of the preparations recommended in this guide will serve you in any crisis situation in a cold climate. You can't make good decisions if you panic, and the Y2K bug is no better reason to panic next winter than possible blizzards or earthquakes were last winter.

Should the Y2K 'earthquake' turn out to be just a tremor, your efforts won't be in vain, since now you'll be prepared for the equivalent of the ice storm.

And we believe that the activity and accomplishment of preparing, in whatever fashion you choose to do so, will lower any anxiety you may feel about the consequences of the Y2K bug.

If anxiety does become a significant problem for you, you may want to consider professional counselling, but in the meantime, we believe that reasonable preparations for potential Y2K problems are your best preparation for peace of mind.

By the time you finish reading this book, you should have your work-plan completed and be ready to go. Good luck in your activities and remember: 'A problem shared is a problem halved!' Talk with your family, friends, and neighbours.

IF YOU LOSE POWER

We tend to forget how much we rely on electricity and a dependable fuel supply, especially in winter.

And we tend to forget how dangerous electrical power and heating fuels can be when mishandled.

Unless you're well prepared, losing power and heat can be a matter of life and death.

Like the rest of this book, this chapter puts safety first. So should you. Make sure that everything you do adds to your safety and comfort, rather than adding to your risks.

Since we live in one of coldest inhabited parts of the world, heat must be our central concern if the power is lost in typical January conditions.

Just consider that on many prairie or arctic nights, the difference between the temperature inside your home and outside may be more than 45°C! The cold can be particularly hard on infants and children, the elderly, and those with disabilities.

If the power goes off, it may because of a short-term, localized problem, or it may be longer term and more widespread. When it comes back on, it may be back for good, or it may still be interrupted by 'rolling' brown-outs or blackouts as the power grid tries to cope with localized failures.

HOW MUCH TIME DO I HAVE?

Don't panic if the power goes off. Keep in mind that even in very cold weather, it will take several hours before a closed living space will become too cold for comfort.

How much time you have before water pipes begin to freeze depends on the outside temperature, the efficiency of your home insulation, and many other factors. As a rough guide, you should figure on the house temperature dropping by about *1°C per hour*.

At that rate, your home might reach the freezing point in about a day, putting anything freezable at risk and making your life uncomfortable. You should consider mounting a thermometer inside your home so that you can easily monitor the indoor temperature beyond the lower range on your thermostat.

At 4°C it's time to drain all the pipes if it looks like the temperature is going to continue to fall. See details below under 'When to Worry About the Plumbing.'

LIVING WITHOUT HEAT: FALLBACK SOLUTIONS

In a prolonged blackout without backup heat, you may *have* to abandon your living space and head for a community shelter. But until that time, you have a good option: you can create a survival space that doesn't rely on an additional heat source. This is one of the most important recommendations of *Weathering Y2K in Canada*. *Everyone* can do it, whether you live in your own house or rented accommodation. And it costs nothing.

HOW TO MAKE A SURVIVAL SPACE IN YOUR HOME

A survival space is very small and heavily insulated to retain warm air. It does not require a heat source.

Move into one room or part of the house that can be closed off and insulated. Your goal will be to insulate this room with readily available household items.

Choose a small room. It should be exposed as little as possible to outside cold, meaning that it should have as little outside wall and as few windows as possible. South-facing windows are best if it has windows, and they should be double-glazed or sealed with clear plastic (most Canadian hardware and building stores sell kits for sealing windows). The room should have draft-free doors that can be left ajar for airflow when necessary.

Then make a shelter within this space, remembering that insulation is critical. If the walls are well insulated, you might pick a corner and use mattresses to form the other walls of your 'nest.' Your body heat alone can keep this space fairly comfortable when it's kept in by insulating materials such as blankets, sleeping bags, towels, duvets, heavy clothing, drapes, and even area rugs. Make sure that you put an extra layer of insulation such as a throw carpet *underneath* you.

Cover the entrance with a blanket that can be drawn aside to permit some ventilation. Stay under the covers if it's really getting cold. Remember: the more bodies in there with you—maybe even including your pets—the snugger it will be.

You can be creative with the construction of your 'room within a room,' using only what you have at hand, although you might want to add a good sleeping bag or two. Remember that conserving heat is easier than making it.

The June 1997 issue of *Consumer Reports* compares sleeping bag features.

When the survival space is in use, avoid moving in and out of it as much as possible. Don't leave the house directly from this room. Let sunlight in if it's direct enough to warm the room; otherwise draw your curtains. In any case, be sure to draw the curtains at night.

OPTIONS

Another option may be open to you if you have an alternative heat source such as a wood stove or kerosene heater. You may be able to keep part of your home at near-normal temperatures. This 'warm space' has many of the characteristics of the survival space but the main difference is size and much-increased ventilation that allows the use of heating devices.

Obviously, a room could serve as the warm space while part of the room could be cocooned into a survival space as long as no heating or cooking device is used.

You may want to consider two warm spaces: one for your living space and another for food, water, and other items that you do not want to freeze, such as house plants. Or you may want to make your one warm space large enough to accommodate all your essentials.

❋ **Reminders**
  ❋ Carbon monoxide (CO) can kill. Buy and use a battery-powered CO detector.
  ❋ Although the idea when keeping warm is to create an area with minimal air movement, ensure that there is some ventilation, since CO detectors may malfunction.

* Don't use candles, gas cookers, supplementary heaters, or other flames in your survival space.
* Buy and use a battery-powered smoke detector.
* Keep your radio, flashlight, fire extinguisher, CO and smoke detectors close to your survival space.

WARM CLOTHING

Whether or not you have backup heating for your home, you should dress warmly indoors to conserve fuel.

Over the past decade there has been a revolution in clothing technology for extreme weather conditions. Polypropylene underwear and layers of 'fleece,' used in conjunction with a water-resistant outer layer like Gortex™, work extremely well to keep you warm. Among their chief benefits is the ability of polypropylene and fleece to wick away moisture and dry quickly. The only natural fibres that come close are wool and cashmere. Cotton is *not* recommended for situations where you need to stay warm.

As you've heard before, dress in layers—sheer bulk is not as important as trapping air between the layers of clothing.

Wear a wool or fleece hat. About fifty per cent of your body-heat loss is from the head. Wrap a scarf around exposed parts of the neck or face, and ensure that your abdomen is well covered. And don't neglect your feet. Maintain a stock of woollen socks (not cotton blend) and keep your feet dry when venturing outside. The same applies to gloves: wool or fleece is best, and your hands will be even warmer if you pull waterproof mittens over your gloves.

HOW TO SAFEGUARD YOUR HOUSE WHEN THERE IS NO POWER

* Get your flashlights. If you've installed battery-powered emergency lights in central or strategic spots, you'll find things more easily. (See 'Light' section.)
* If your power goes out, first check to see whether other houses or buildings in your neighbourhood have power.

    If houses outside your immediate neighbourhood have lights, this could be a very localized problem, and your best resource could be your neighbours.

* Tune in your battery-powered radio and listen for local information.
* Shut off and unplug sensitive electronic equipment such as your TV, VCR, audio equipment, computer hardware, and appliances with electric motors such as your fridge. When power is restored, or if there are recurring brown-outs, there is a good chance of damaging power 'spikes' on the line.
* Some power companies and Emergency Preparedness Canada advise homeowners to turn off the main breaker or remove the main power fuse. This protects electric equipment from power spikes and avoids danger for the homeowner while working around the furnace, hot water heater, appliances, and plumbing (which some homes use as an electrical ground).

   You can choose to leave the breaker off once you've finished work on the electric appliances and plumbing, or to switch the breaker back on and leave one light on (perhaps in your warm room) so you'll know when the power is restored.
* You should already have checked where the main gas shut-off valve is your home. In older houses, the gas valve is often inside whereas in newer homes, the shut-off valve is usually outside next to the meter.
* Don't turn off your natural gas unless advised by local authorities. The gas companies expect to maintain pressure, and your pilot light should not be extinguished by a power failure. Nearly all furnaces have a regulator that cuts off the gas if the pressure is too high or low, or if the pilot light is out: no flame—no gas.
* The same applies to your hot water heater. Note that some older gas kitchen ranges are not equipped with an auto shut-off, so if the pilot light is extinguished, low pressure gas can still continue to flow. Check your owner's manual or your appliance well ahead of time.
* As a precaution, we recommend that you still check the furnace area and kitchen from time to time. If you smell gas or oil, put safety first. If the problem is very slight, turn off

the fuel supply and ventilate the area. Otherwise get people out first, and then call for professional assistance. Remember that even a static electricity spark can ignite trapped or pooled vapours.

❊ If you have a backup heating unit, get it going before your home gets too cold.

WHEN TO WORRY ABOUT THE PLUMBING

Well ahead of time, you should find your main shut-off valve and your system drain valve and make sure that each works properly. These are normally near where your water supply enters your home.

❊ When the temperature in your house approaches 4°C and the outside temperature is lower still, or if you decide to vacate all or part of the home, it's time to drain your inside water pipes and appliances so that pipes don't burst. Four degrees Celsius is when water starts to expand. Better a little sooner than a little later. At -1°C, the pipes will quickly reach the point of no return.

❊ You should already have disconnected garden hoses and drained water to vulnerable pipes leading to outside taps.

❊ Shut off the main shut-off valve where it enters the house. Protect the valve, inlet pipe, and meter or pump with blankets or some other insulating material. (Plastic sheeting has almost no insulating value.)

❊ If your home is heated with a warm-water system, turn off any electrical connection, open the radiant equipment drain valves, and then drain the furnace.

❊ Drain the water from your plumbing system. Starting at the top of the house, open all taps and flush toilets several times. Use a large sponge or a turkey baster to empty water in the toilet tanks.

❊ Go to the basement and open the drain valve.

❊ If you have an electric water heater, ensure that the circuit breaker for the heater is off, and remove the water heater's electrical connection before draining the tank. Attach a hose to the tank drain valve and run it to the basement floor drain, or put a bucket under the valve as you empty the tank.

* Make sure that you have disconnected the power from all electrical appliances. Check the instruction book for your dishwasher, and drain any water remaining in it. Do the same with your washing machine, the humidifier on your furnace, and the refrigerator if it has a water connection for icemaking or a cold-water dispenser.

  Don't worry about small amounts of water trapped in horizontal pipes. Add a cup of plumbers' antifreeze to each toilet bowl. Squirt antifreeze into every faucet and add about half a cup of antifreeze to each drain, not forgetting shower stalls, bath, and floor drains. This should prevent residual water from freezing and causing damage.
* Listen to your radio for more-detailed local advice and instructions.
* Never attempt to thaw pipes with any kind of flame device. You risk setting your house on fire.

AFTER THE POWER RETURNS

* If the temperature in your home is above freezing, turn on the water supply. If it's below freezing, warm it thoroughly before turning the water back on.
* Close lowest taps first and allow air to escape from upper taps.
* Check for any leaks caused by freezing and for any drains still blocked by freezing and, if necessary, shut the water supply off again until drains thaw, or until you or a professional have made the necessary repairs to prevent water damage to your house. Check for leaks frequently over the next several days.
* Turn off your standby generator.
* Ensure that the hot water heater is filled with water before reconnecting the power.
* Rinse out dishwasher and washing machine if necessary.
* Switch the main circuit breaker back on if it's still off, or replace the main fuse.
* Warm house slightly above normal temperature for a few hours to allow it to dry thoroughly.

❋ Further notes about the plumbing system

❋ Indoor sprinkler systems, hot tubs, and indoor whirlpools may require special handling. Read your manuals and, if necessary, get advice from the dealer or a professional.

❋ Your hot water heater probably holds a hundred litres or more. Because the tank is insulated, it will hold its temperature for more than 24 hours. This warm water is a valuable resource that you may want to conserve for a time when you're draining your system. If you decide not to drain your tank immediately, be sure to monitor it carefully as it cools so that you don't risk freezing damage.

❋ Plumbing tool kit

You will need some tools and other items to complete these tasks:

❋ wrenches or adjustable pliers

❋ a short length of plastic hose for syphoning

❋ a short length of garden hose with a female connector for draining your system and your hot-water tank

❋ a sturdy bucket

❋ a squirt bottle or turkey baster

❋ four litres of RV (recreation vehicle) or plumbers' antifreeze. Only use *non-toxic* antifreeze; automotive antifreeze is poisonous and must not enter sewer systems or your drinking water. Four litres of Prestone™ RV −50° antifreeze, for example, cost about $6.00 at hardware stores such as Canadian Tire and Revy.

When you consider the risk of split pipes spewing water behind walls and over ceilings, it's worth taking these plumbing precautions seriously. If you've done it right, all you'll need to be concerned about when your house is warm again is that the coloured antifreeze is thoroughly flushed away.

HEAT, COOKING, AND LIGHT

The alternative devices for heat, cooking, and light have so much in common that we are going to deal with them together. First we need to ensure that you are aware of the safety issues that surround the use of these devices in crisis situations. Bear with us while we cover the dangers of all the alternatives that we present in the later sections.

HAZARDS OF FUELS AND USING FLAME DEVICES

Many temporary solutions for heating, light, and cooking involve flame. This section alerts you to the risks involved, and offers advice for minimizing the danger.

In winter survival conditions, you stand a much greater chance of death or illness from carbon monoxide poisoning than from cold or fire, although cold and fire both pose serious risks.

If you take nothing else from this guide, remember our warnings about the deadly risks of burning any fuel: wood, kerosene, propane, charcoal, coal, and even natural gas.

Halfway through the ice storm of 1998, fifteen fatalities due to carbon monoxide poisoning had already been recorded. Many of the people who used fireplaces on a continual basis suffered some degree of carbon monoxide poisoning, as did others who tried to heat their homes with makeshift devices. White gas stoves, barbecues, and propane stoves were the cause of many carbon monoxide casualties.

The ice storm was unusual but the risk is all too common: in the U.S., carbon monoxide causes more accidental poisonings than any other chemical substance.

CAUTIONS & WARNINGS FOR YOUR BACKUP PLANNING

* Be aware of inherent risks in your choices of backup measures. Read and follow manufacturers' directions on everything you use, and when uncertain, err on the side of safety.
* Do not use your gas or electric stove as an alternative heat source.
* Use devices certified by the Canadian Standards Association or Canadian Gas Association.
* Store fuels safely and legally.
* Buy at least one fire extinguisher designed for all types of fires (rated 'ABC').
* Buy and install one or more battery-operated carbon monoxide detectors.
* Do not, under any circumstance, use any flame device—including candles—inside a tightly sealed living area.
* Never sleep with an unvented combustion device—you could die of carbon monoxide poisoning or lack of air.
* Be aware that, in a tightly sealed home, a strong draft up one chimney, such as your fireplace, can cause a backdraft down another chimney, such as a propane heater. Make sure that your combustion heaters have a fresh-air supply (open a window or door a crack if necessary) to avoid sucking combustion products back into your home.
* Fuels vary in how easy they are to store safely, and how dangerous they are to use indoors.
* Treat anything that burns fuel, and the fuels themselves, as dangerous.

Not many fossil-fuel devices are safe to use inside a home, but we discuss below the products on the market, along with their merits and shortcomings.

Permanent, installed heaters with proper venting or chimneys provide your safest backup, but, confronted by difficult choices, many planners may be attracted by the cost-effectiveness of portable solutions.

Our recommendations are few, while our warnings are many and on many levels. Our key recommendation: take responsibility for your own safety, and take it seriously.

WHAT TO LOOK FOR IN A CO DETECTOR

Your best investment of the year: buy a battery-powered carbon monoxide detector with a digital read-out.

Carbon monoxide (CO) is often called 'the silent killer' because none of our senses can detect it. Even if you're taking proper precautions with ventilation and venting, you need a detector.

Some recent tests suggest that not all carbon monoxide detectors perform to specifications, even some of those marked as Canadian Standards Association or Underwriters Laboratories certified. Some are overly sensitive according to some tests, and others fail to sound at dangerous levels of carbon monoxide. Check the most recent information available before buying a detector.

If you already have a detector, still check the most recent information available. Relying on a malfunctioning detector may actually increase your risk.

The Consumer Product Safety Commission recommends at least one detector per household, installed near the sleeping area. You may wish to have more than one detector in your home.

The November 1996 issue of *Consumer Reports* has a feature report on carbon monoxide detectors with an update in the January 1998 issue. Consult it at your library or online at:

http://www.consumerreports.org

Consumer Reports highly recommends two battery-powered models with digital read-out: AIM™ Safety SAS-696D, and Nighthawk™ KN-COPP-B.

We have seen new First Alert™ and Nighthawk™ models at Revy, London Drugs, and other stores. Price: about $54–$70.

First Alert™ can be reached at 1-800-722-1938, and Nighthawk™ at 1-800-880-6788.

FIRE EXTINGUISHERS

Whether or not you're concerned about the weather or the Y2K risk, you should acquire this essential home-safety item. There are a few things to take into account when making the decision on what type of fire extinguisher to get.

Fire extinguishers are rated according to the types of fires they will extinguish:

A combustibles such as papers, rags, wood
B oil, gasoline, grease, solvents
C electrical fires.

We recommend purchasing at least one extinguisher rated 'ABC,' meaning that it works on all types of fires. Size means endurance: we recommend at least two kilograms of chemical. The rating you see on packaging may include numbers: for example 2-A:10-B, C. These numbers signify what kinds of fires the extinguisher is 'weighted' towards: in this case mainly 'B' fires, then 'A,' then 'C,' which reflects the types of fires you are most likely to encounter in your home.

Read the instructions as soon as you unpack the fire extinguisher—not in the crucial seconds that you may have to bring a fire quickly under control. Your actions by then should be automatic.

Prices for commonly available fire extinguishers start at about $50 for 2.27 kg models.

SMOKE DETECTORS

We hope that your home already has these lifesavers installed on every level and in every bedroom. Keep in mind their power requirements—replace those that are dependent on household current unless they have battery backup, and regularly test your detectors' battery status. Replace all the batteries before the coming winter.

There are two basic types of smoke detectors now on the market: 'photoelectric sensor' and 'ionisation sensor.' The former is highly effective at detecting slow, smouldering fires; the latter is best for early detection of fast, flaming fires.

Examples of battery models:

* First Alert™ SA 301A. Incorporates both types of sensors. Price: about $35.
* First Alert™ SA 76RDCA. Basic protection. Price: about $12.

FUEL CONSIDERATIONS

Is it safe to use liquid fuels and propane or butane gas in your home?

The answer to this question has a major bearing on what, if any, supplementary heat, cooking, and light devices you use in your home.

While some fuels burn more efficiently than others, *all* fuels produce some carbon monoxide and in a poorly ventilated space, that can be lethal. Do not, under any circumstance, use any flame device—including candles—inside a tightly sealed living area.

Fuels also vary in how easy they are to store safely, and how dangerous they are to use indoors. Review the information below before making any decisions about what fuel-burning devices you will plan to use as backups, and how you will use them.

❄ Kerosene

Kerosene is variously known as 'Range Oil No. 1,' coal oil, paraffin, and lamp oil, which is often scented. The only type that you should consider using in heaters, stoves, or lanterns is the highly refined No. 1–K available at some camping and hardware stores.

The big advantage of kerosene is that it is a non-explosive fuel, classed as a combustible with an ignition point above 40° C by the National Fire Code of Canada.

Kerosene is quite forgiving and has a long shelf life. It can be transported safely, purchased in bulk, and decanted to smaller containers. Kerosene consumes oxygen and produces some CO when it burns so you must ensure that fresh air circulates: at the very least crack open a window or door.

❄ Propane/butane

Propane is a hotter-burning fuel than the natural gas that is used in many furnaces and kitchen ranges. Normally propane devices are supplied by large tanks installed outside the home. Propane devices produce non-poisonous carbon dioxide as they burn—no problem there—but they often consume large amounts of oxygen and, like all carbon fuels, produce traces of carbon monoxide.

Portable propane heaters, stoves, and lights that use disposable canisters are not recommended for use inside a dwelling. If you decide to use a propane device inside a garage, cottage, or ice fishing shelter, think ventilation! Always open a door or window to the outside when using a propane heating, cooking, or light device.

Additional warnings on propane:

* Avoid leaks—propane is heavier than air and can accumulate in explosive pockets in low spots in the home or storage buildings.
* Propane tanks, full or 'empty,' should be stored in a well-ventilated structure unconnected to your dwelling, preferably one that locks. Most local authorities strictly control propane storage. We advise that you check with your local propane dealer, the fire department, and your landlord.

❄ Natural gas

If you're using natural gas (normally 80%–95% methane by volume), the same cautions apply as for propane. Although natural gas is lighter than air and does not pool in the same way as propane, it is toxic and explosive in confined spaces.

❄ Gasoline

If you're running a gasoline generator, never forget just how dangerous gasoline is. Never store or pour gasoline inside your house. Use only containers designed for storing gasoline. Unless you're lucky enough to be on a farm with a proper storage tank, the only safe storage is in a well-ventilated shed or a ventilated, detached garage.

Gasoline deteriorates if stored for longer than six months, so if you're buying as a Y2K precaution, purchase gasoline supplies in the last months of 1999.

❄ White gas

Also known as 'naphtha,' 'camp fuel,' 'Coleman™ fuel,' 'Coleman™ gas,' and 'Hi Test.'

For good reason, white gas is at the bottom of our list of fuels for heating and cooking. Although quite a few portable cook stoves (including the venerable Coleman™ two-burner) and lanterns use this fuel, it is extremely volatile and tricky to use safely, espe-cially indoors. We have witnessed several explosive incidents in camps where people have tried to refill hot stoves, or have had major flaring during attempts to prime stoves and lamps. Similar incidents inside a home could be disastrous. Because white gas stoves and lanterns often burn inefficiently, copious amounts of CO can be given off. So if you're going to use white gas devices, use them cautiously, and use them outdoors.

HEAT

ALTERNATIVE HEAT SOURCES

Virtually all natural gas and oil furnaces shut down automatically if there is no power for the furnace blower and thermostat con-trols. Without mains electricity, the only way to fire up a dead furnace is with a backup generator—and that's a complicated sce-nario covered in a later section.

Important questions include:

* Are there safe options?
* Do you want to invest in a backup heat source?
* Should this heat source be portable, semi-permanent, or permanent?
* What can you afford?

The section below starts with the simplest alternative heat sources with the lowest level of financial commitment, and works up to semi-permanent or permanent installations. You can stop reading when you've reached your comfort level!

* **Choosing a heat source**
  Your choice depends on several factors:

  * Do you rent or own your home? Some of the suggestions that follow would be impractical for those living in rented accommodation; your landlord or municipal regulations may not permit use of portable heaters.

* Do you have the ability and permission to modify your living space? Your choices may be limited if you live in a high-rise building.
* Fire safety and carbon monoxide considerations should be your most important guiding principles. No situation is worth compromising your family's safety or others in the same building.
* What do you want to spend? Not everyone has a lot of money to spend on a backup system you may never use.
* What's available? You may want to consider more than one option, since you may find that some of the more popular devices are in short supply.

We asked gas and electricity technical advisors what they would recommend for home heating in the event of power failure, and their response was to discourage anything that vents into the living space because of the potential problems of exhaust gases, especially carbon monoxide.

With that advice in mind, you still have a few options:

* a kerosene heater (used with great caution and never while sleeping)
* a wall-mounted propane heater, vented to the outside
* a generator (note the caveats regarding fuel, noise, and safe connection)
* a wood stove, professionally installed
* a free-standing natural gas or propane stove, vented to the outside
* a fireplace insert, wood or propane, professionally installed.

PORTABLE HEATERS

Our 'jury' is still in session on this class of heaters. As we've heard from heating professionals, you shouldn't vent the products of combustion directly into your living space. You risk carbon monoxide poisoning.

But, if used with *extreme caution*, in circumstances where there is a flow of *fresh air*, some kerosene heaters might be a reasonable option for keeping your house warm if the power goes out. You must, however, take the necessary steps to use them properly and safely.

We recommend against using portable propane heaters. No one we've talked to will go on record to endorse the use of portable propane heaters in occupied homes.

❄ Portable kerosene heaters

Price: roughly $300–$500

The manufacturers of kerosene heaters claim that portable kerosene heaters certified by the Canadian Standards Association are among the safest of all supplemental heating devices, particularly when compared to white gas and propane heaters. They have been used for years in shops and garages and on farms.

With that said, we note that a popular model comes with a prominent warning including the following excerpts:

> *Risk of Air Pollution. Use only in well-ventilated areas.... [I]f the heater is used in a small room...the doors to adjacent rooms should be left open or a window to the outside should be opened at least 1 inch.... [D]o not use the heater in a bathroom or any other small room with the door closed.*

KERO-SUN™ Omni 105 Operating Instructions

Keeping such warnings in mind, kerosene heaters may prove to be a popular solution for area warmth. Models include convection heaters meant for large areas, and radiant heaters designed to target heat in a specific direction.

Examples include:

❄ The Kero-Sun™ CTK-25 Sunsprite convection room heater generates up to 8,500 BTU (suitable for about 350 sq. ft.). It operates for 23 to 33 hours on 6 litres of fuel. Price: about $290.

❄ The Kero-Sun™ Omni 105 convection heater, generating up to 15,000 BTU (suitable for about 600 sq. ft.). It operates for about 18 to 24 hours on 7.4 litres of fuel. Price: about $390.

Some models are already in short supply so we would recommend checking your local hardware store or camping store as soon as possible; alternatively, you can order direct from Kero-Sun at **www.kerosun.com**

❄ Notes on kerosene

While kerosene is not explosive, the manufacturers recommend that you fill the heater outside to avoid possible spills. You can expect slight odours on lighting and extinguishing the heater, so you might also consider doing those operations outside. But do not move an unit that is flaring.

Highly refined 1-K standard kerosene sells in containers of various sizes:

1 litre	about $2.50
4 litre	about $7.50
9.46 litre	about $19.00
20 litre	about $35.00

With an 8,500 BTU heating unit, a 20-litre pail should last 4 to 5 days.

And remember, while kerosene is relatively safe in comparison to other fuels for portable heaters, you should still treat it as hazardous and take appropriate precautions if you're using it for backup heat.

❄ Note on portable propane heaters

Although relatively clean burning, propane heaters produce some CO and consume copious amounts of oxygen. They should only be used in very well-ventilated settings such as ice-fishing huts, drafty cabins, open garages, or football games. Deaths have been attributed to misuse of this kind of heater: a recent Alberta multiple fatality was due to campers sleeping with a propane heater in a sealed tent.

As we've described above, we believe that propane used with portable devices brings unacceptable risks into an occupied home. We therefore recommend against using portable propane heaters as backup heat for your home, but we still encourage you to read the following section on wall-mounted propane heaters.

WALL-MOUNTED PROPANE OR NATURAL GAS HEATERS

Price: from about $900 plus installation

These compact wall furnaces (sometimes called 'garage heaters') are supplied by propane tanks outside your home or from a natural gas line. They may be an answer for relatively open spaces in a garage, trailer unit, or cottage where it's possible to cut holes through the exterior wall for venting and for the fuel supply line.

These furnaces do not require fans. Their gravity-fed gas flame is completely sealed inside a combustion chamber that vents combustion products directly to the outside. This allows you to keep windows and doors closed. In moderate conditions, a 50-lb. propane cylinder will provide heat for about one week. If you have a natural gas heater, remember that gas supplies may fail.

Installation includes gas-fitting and must conform to local codes, so you should hire a professional.

Example: Debonair Direct Vent Wall Furnaces made by Williams, ph: 1-800-266-0993.

GENERATORS

Price: $600–$2,100 plus hookup to your home

We have mixed feelings about generators. They're expensive, noisy, big consumers of explosive fuel, of very limited capacity compared to your line supply, and risky if they're not hooked up by experts. On the other hand, they can enable you to use some of the many electrical devices in your home that will otherwise just take up space until the power comes back on.

Depending on the size of the generator, it can power a few lights, allow intermittent use of electrical appliances, and—for a limited time—supply power to your heating system.

But for most people in urban areas, especially those in rented accommodation or in apartments, a generator is not a realistic option.

❄ **Generator warnings**
  ❄ Never, under any circumstances, run or refuel the generator anywhere inside your home or apartment, or even in an open garage.

* Expect to be unpopular in your neighbourhood if you run a generator; all except the most expensive tend to be noisy and temperamental.
* The voltage supplied by your generator may vary considerably, producing spikes that could damage electronic equipment or appliances.
* Storing, transporting, and pouring large quantities of gasoline is dangerous. Transport and store your supply in small quantities in proper fuel containers. Store in locked, ventilated spaces well away from your home. Make sure that you can easily lift and pour from your storage containers, so that you can fill the generator tank without injuring yourself or spilling fuel.
* Light-gauge extension cords carrying power over long distances will overheat, creating a risk of fire and shock. If you're connecting your generator to devices or your home with extension cords, check technical specifications when determining what gauge cords to use. Typically, for a 1,000-watt generator 60 m from the appliances, you will need a 12-gauge cable.

Generators quickly become very scarce when people face the risk or the reality of power going out. They are in short supply in some areas already. If you want but can't get one, don't worry: you can plan to get along without one.

You might want to seriously consider whether you will have a use for a generator once millennium worries are behind us—perhaps for recreation, building projects, or as a hedge against future weather emergencies. And there is always the re-sale market in rural communities.

* Is it possible, and safe, to link a generator into my home electrical system?

If you decide to use a portable generator, you can connect it directly to small appliances with a heavy-duty extension cord. Check the technical specifications for both the generator and the cord before making a connection.

On the other hand, you must get a permit and hire a qualified electrician to connect the generator directly to your home's electrical panel.

Very serious safety concerns exist when you connect a generator to a home's electrical panel. If connected incorrectly, a lethal current can feed back to pole-mounted transformers, endangering service people who don't expect to find live lines in a blacked-out area. You also risk explosion with an improperly connected generator when the power comes back on.

❄ What size is useful?

A five-kilowatt model running twelve hours a day will need about thirty-four litres of gasoline per day. A 2.5-kilowatt generator will use a little less than half that fuel.

We believe that for most people, a 2.5-kilowatt generator is at the low end of practicality: it could power, for example, a furnace fan, two appliances, and two lights. A 2.5-kilowatt generator won't power 220V appliances, such as an electric stove, an electric dryer, or electric heating.

Example

2 x 60 watt lamps	120 watts
Microwave	1,200 watts
Crock Pot	70 watts
Gas or oil furnace fan (including start-up current)	510 watts
TOTAL	1,900 watts

It may be worth checking Canadian Tire and Honda dealers for available stock. A new model on the market—the Honda EU3000isCAO— is in particular demand because of its fuel efficiency and relatively quiet operation. Specs: 3000 watts, 13-litre tank capacity, runs 20 hours on 25 per cent load or 7.3 hours on full load. Price: about $2,200.

A more economical model, the Honda EZ2500XKIC, costs about $1,100. Its tank capacity is only 3.7 litres, which lasts for about 2.8 hours at full load.

❄ **Further reading**
Other good sources of online information on generators include:

* ❄ 'Select the Right Portable Generator'
  http://www.ext.vt.edu/pubs/disaster/490-303/490-303.html
* ❄ 'Everything you wanted to know about power and generators'
  http://theepicenter.com/tow1230.html

WOOD STOVES

Price: $200–$2,500 plus cost of installation (about $800–1400) One of the important lessons of the 1998 ice storm is that the oldest and most basic fuel—wood—was still an important source of heat when the lines were down. The wood stove looks like an ideal solution for backup heat this winter, but you should be aware that there are complications and additional costs, including the wood itself.

While it is possible to buy an inexpensive basic model, the buyer then faces substantial installation costs: typically about $800 for a single-storey bungalow where the stove is installed on the main floor.

Since hot air rises, it would seem logical to install a wood stove in the basement as a backup for gas or oil-fed furnaces, since this would most likely solve the water-pipe problem. However, Natural Resources Canada recommends that you put the wood stove 'where you live' since hot air does not circulate through the rest of the house very effectively without forced ventilation, and your basement could overheat.

If you have been thinking of installing a wood stove for its looks and functionality, it might make sense to do it earlier rather than later this year. Some homeowners may choose attractive designer models and install them on the main floor where they add to, rather than detract from, the market value of the home. There will be life after Y2K!

Wherever you locate the stove, you should use qualified tradesmen to install the high-temperature, double-walled flues and chimney. This is not a do-it-yourself job.

Basic stove models are available in most hardware and building stores, and many communities have stores specializing in stove and fireplaces.

Canadian Tire, for example, stocks a range of basic models rated for 1,000–1,600 sq. ft. areas, made by Century Heating, Ontario, ph: 416-798-2800. Price: about $200–$550.

Regency, a Canadian manufacturer, has an extensive range of fireplace products, including a recommended model: F2100 Medium, which has a 70,000 BTU rating for 1,000–2,200 sq. feet. Price: about $900 plus installation. An informative site for the range of Regency products is at

www.regency-fire.com

One high-end example: Dutchwest Federal Convection Heater suitable for 700–1,400 sq. ft. with a maximum heat output of 35,000 BTUs and a burn time of up to 8 hours. Price: about $1,200 plus installation. The web site for the manufacturer can be found at

http://www.majesticproducts.com/products/index.htm

Note: an excellent, free publication, *A Guide to Residential Wood Heating*, is available from Natural Resources Canada. This guide includes a very full discussion on wood heating options, safety, and efficiency. Phone: 1-800-387-2000 or download free at

http://www.nrcan.gc.ca/es/erb/reed/wood/index.html

❄ Notes on wood

Different wood types have different heating characteristics. You get more heat from a hardwood than a softwood: hickory produces almost twice the heat of aspen, white pine, or black spruce, for example. Although hardwoods such as beech and maple are plentiful in parts of Canada, there are areas where softwoods are the only available fuel woods.

Wood from conifers, such as pine, fir, and spruce, is high in resins that will produce creosote, so if you're burning these woods you'll need to be even more wary about chimney fires.

Most firewood will need about six months of air drying to burn well, so buy early.

An excellent discussion on *Purchasing and Preparing Your Fuel Supply* can be found at

http://www.nrcan.gc.ca/es/erb/reed/wood/11_e.html

WOOD AND PROPANE FIREPLACE INSERTS

Price: $1,700–$1,800, plus about $450 for installation
Fireplace inserts convert existing wood-burning fireplaces to more efficient operation. Both wood and propane gas inserts are available.

A sealed-combustion fireplace insert allows you to keep the warm air where you want it: in your home instead of up the chimney. It has features that a conventional fireplace lacks including a tight-fitting gasketed door and a heat exchanger. Most inserts have glass doors so you can see the flames.

Some gas inserts provide heat outputs ranging from 10,000 to 40,000 BTUs and burn at around 80 per cent efficiency. It should be possible to maintain a reasonably comfortable temperature in a 1,000–2,000 sq. ft. area using a fireplace insert such as the Regency I2100 Medium.

Note that some models have electric blowers or fans that may reduce their usefulness during a power outage. Contact a fireplace retailer for further information.

FREE-STANDING NATURAL GAS AND PROPANE STOVES

Well-designed models are available from Canadian manufacturers, including the Regency line, which use either natural gas or liquid-propane burners. You may need to special order the propane models, but later in 1999, conversion kits will be available.

If you decide to get a natural gas or propane stove, check whether it has electronic ignition. If the power goes out, you could be out of luck. Regency models are able to operate without electricity by using a gas-heated thermopile that generates enough power to control the heating device. The electric fan may not work, but you would still have radiant heat.

Examples: Regency Ultimate U45 38,000 BTU. Price: about $1,800, plus $445 for installation.

Canadian-made 35,000 BTU Kozi Model DV1 Direct Vent

Gas Stove, from APR Industries Ltd., Winnipeg. Price: about $1,500 plus installation.

❋ Note on natural gas vs propane

One could make the case that a natural gas, free-standing stove is a better buy, because there is no need to refill propane cylinders. But it is a judgement call whether natural gas will continue to flow if the power grid experiences Y2K problems. Like so many aspects of the Y2K problem, the public may not get definitive answers until late 1999.

SOLAR AND BATTERY SYSTEMS

Price: $2,000 and up

If you're interested in permanently minimizing your dependence on your local utility company, you may want to consider more exotic and expensive solutions. You should be aware, however, that these are not quick remedies, and they're not for the mechanically or electrically challenged.

You can buy battery systems, recharged from the power grid, with 1,800 watts of short-term usable stored electrical power, starting at about $2,000.

Stand-alone systems using solar panels supplying up to 4,800 watts of usable energy start at about $10,000.

A good online site is at

http://www.mrsolar.com/Solarkitcabin.html

❋ Notes on fireplaces

In cold weather, your open fireplace is good for psychological comfort but that's about it.

Unless your fireplace is of advanced design, the fire actually draws more heat up the chimney than it delivers to your house. A conventional fireplace has no effective way to transfer heat from the fire into the room, so most of the heat rushes up the chimney, sucking air with it. You *can't* heat the whole house with an open fireplace.

In the 1998 ice storm, while many families successfully used wood-burning stoves and fireplaces, a few lost their homes because of unsafe conditions and their own inexperience. Wet wood, make-shift wood stove installations, and long-term heating with fireplaces not designed for continuous use increase the risk of a house fire.

We do not recommend the constant use of ordinary fireplaces as your emergency heat source.

If you do use your fireplace more than usual, keep the fire small rather than building it up in a counter-productive attempt to warm areas in your home beyond its reach. Avoiding intense fires also helps you avoid overheating in the chimney or surrounding woodwork, which can cause house fires.

Smoky fires resulting from moist wood or poor ventilation can produce highly flammable deposits in your chimney. Enough creosote to sustain a dangerous chimney fire can be deposited in just a few days of continuous fireplace use.

Our strong recommendation: get either a licensed installer or an accredited chimney sweep to check that your chimney is clear of creosote and obstructions such as bird nests, well before the end of 1999.

❄ Don't use your gas kitchen stove for heat

Exercise caution. When your house starts getting cold, you might be tempted to rely on your stove, but it's not a good idea.

Your natural gas kitchen stove may still work when the power is down, but remember that your range hood fan will not, so combustion products will not be vented to the outdoors. Normally, gas ranges only emit nitrogen oxides, but if the stove malfunctions or if you have poor ventilation, carbon monoxide can build up.

A gas range can be used intermittently with adequate ventilation but it should *not* be used continuously to heat a room.

If for some reason electrical power is available but not fuel for your furnace, similar advice holds for your electric kitchen range: it is not designed to heat a room. If you use it for that purpose, you risk damage to your appliance, and fire.

COOKING

SAFETY FIRST

* If you're cooking on a portable stove or a camp stove, cook outside. Next best would likely be on the stove under a fire-resistant range hood, with a window open and a fire extinguisher handy.

* If you do bring your portable stove indoors, which we recommend against, you need to take every precaution with fuels and flames for your own safety and the safety of your family.

* Never refuel your stove indoors, and never start your stove indoors. Spilled fuel (especially on a hot stove) and flare-ups on starting create serious fire hazards.

* Changing disposable fuel canisters *must* be done outside.

* Store your fuel safely in proper containers.

* Ensure that you have adequate ventilation in the area where you're cooking.

* Keep a fire-extinguisher handy, preferably at least a 2.27 kg 'ABC' type that works on all kinds of fires. Make sure that everyone in your home knows how to use the fire extinguisher.

* Always have someone tending any open flame, whether it's for cooking, light, or heat.

* Never cook in your tightly sealed survival space. Again: cooking outdoors is the safest option, though it may be a little cool for the cook.

* Install a battery-operated carbon monoxide detector at locations recommended by the manufacturer, but don't rely on the detector alone. Make sure you have adequate ventilation.

* If you're running an electrical generator, you can cook indoors on low-draw electrical appliances, such as electric frying pans and crock-pots, with much less risk than you run by cooking with a flame stove indoors.

* If you live in an apartment or a high rise, you will have trouble cooking outside unless the apartment has a balcony. We recommendation that you check with your building superintendent or owner about what may be permissible.

COOKING ALTERNATIVES

If you grew up using a wood cook-stove, or if you're a camper, you have an advantage when the power goes out. Outdoor folks who can talk knowledgeably about Whisperlites™ and the advantages of butane over kerosene at high altitude will be right at home cooking on their portable stoves outside their back door or on their balcony.

If you've installed a wood heating stove with a flat cooking surface on top, you've got a great solution that won't demand that you learn the lost techniques of cooking with a wood cook-stove. While heating stoves rarely get hot enough on top to bring water to a rolling boil, they're ideal for slow-cooking crock-pot recipes, soups, and stews, and they make it easier to renew your water supply by melting snow or ice. You might want to ask old-timers to share their stovetop recipes and expertise.

We describe some of your options in detail below.

❋ Outdoor barbecues and camp stoves

Portable propane barbecues offer a relatively low-cost, versatile solution. For example:

The Weber Gas Go-Anywhere Grill™ operates on a 465 g (16.4 oz.) disposable propane cylinder, or it can be adapted to run off a larger, refillable tank. Price: about $87.

The Big Boy™ Portable BBQ has fold-up legs, operates on small disposable cylinders, and can be adapted for large propane tanks. Price: about $104.

Consumer Reports ranks the following larger propane barbecues highly in its June 1998 issue: the Coleman Powerhouse Plus™, the Thermos™ Texas Grill™, and the Sunbeam™ Grillmaster™. These models all cost over $235.

❋ Emergency stoves

Several solid-fuel models are on the market that use Hexamine or jellied Sterno™. They have some good features: simplicity, a high flashpoint, no priming or pressure required, and they're cheap (about $7). But think of these as strictly stopgaps for a hot cup of soup when nothing else is available. Only use these stoves outdoors—they produce considerable carbon monoxide.

❄ Portable stoves

You can choose between two main types of portable stoves: those that burn fuel in pre-filled metal canisters, usually a blend of propane and butane, and those that burn liquid fuel such as white gas or kerosene.

The range of models on the market is enormous. We've highlighted a few that you might find at your local camping store or hardware store. Talk about your plans with a salesperson who knows the products.

White gas stoves are intended for outdoor use. If you use one, take it outdoors.

Liquid models include:

* The Optimus Explorer™, a well-designed and rugged model that uses a variety of fuels including kerosene. Field-tested in cold weather by experts: the Swedish Army. Kerosene stoves, while a little messy and smelly on start-up, present little risk of explosion. Price: about $135.
* Coleman™ white-gas two-burner has long been the Canadian camp-out favourite. While this model is *not* recommended for indoor use because it burns hazardous white gas and produces carbon monoxide, this old standby can serve you well outside.

Canister stoves also belong outdoors, but they present fewer risks indoors than white gas stoves do. We recommend that you read 'Safety first' above before considering indoor use.

Canister models include:

* The Coleman™ Propane Two-burner Stove, which we recommend. This much-improved and safer version of the white gas stove is very simple to use, much like a natural gas kitchen range. It has two 10,000 BTU burners and boils a litre of water in less than five minutes. Price: about $65 to $80, plus about $3 per gas cylinder. One 465 g cylinder of propane will last up to 4.5 hours. You can also purchase the

fittings needed to connect the stove to a larger, refillable tank. Cost for an empty bulk tank is around $38.

A larger version has two 10,000 BTU burners plus a very hot 15,000 BTU side burner. Price: about $152.

The Coleman™ Xpedition™ Stove puts out lots of cooking power through two independently regulated burners. It uses 300 g cartridges that last about 42 minutes and cost about $5, so it's not cost-effective for the long haul. Price: about $110.

❋ The Turbo 270 Stove, a simple and dependable design that uses butane/propane cartridges. Price: about $37 plus $5 for each cartridge.

❋ The Camping Gaz Micro Bleuet™, a light model often used by backpackers. This popular stove has a potential stability problem, and changing canisters *must* be done outside. Price: about $44.

❋ **Further reading**

The Coleman Company can be reached at 1-800-835-3278.

Contact Optimus at 1-800-543-9124 or at

http://www.optimus.se/

A very good source of independent reviews on stoves may be found at

http://www.gearreview.com/stovereview98.asp

LIGHT

WARNING

Unsafe use of flame lights, such as oil lamps and candles, causes many home fires. Open flames are just plain dangerous in a confined living space, especially if you're not really used to living with them. Burning anything—including candles—also produces carbon monoxide, so flame lights should only be used in well-ventilated areas.

What does that leave you with?

NON-FLAME LIGHTS

The only non-flame solutions are battery or generator flashlights and cyalume sticks, available at some sports stores for about $2.70 each. When activated—which involves twisting or bending them to mix their chemicals—these disposable lights glow for about 8 hours. Check the package: some work poorly at very low temperatures, and there are cold weather versions that provide light for shorter periods.

Cyalume is safe to use, even in a potentially explosive situation such as checking for gas leaks.

❊ Flashlights

A flashlight is not as safe as a cyalume light stick: the tiny spark produced when you turn on a flashlight may ignite explosive vapours. But battery-powered flashlights and lanterns are the short-term lights of choice for situations where you have no power.

They come with many convenient design features. For example, you can buy models that clip to belts, hats, or jackets. A good head-mounted light such as those used by mountaineers and spelunkers leaves your hands free to work or turn pages. Petzl Zoom™ is the standard by which other head-mounted lights are judged. Price: about $38. Coleman™ markets a waterproof headlight that offers very good value at about $7.

We highly recommend the often-copied slimline Mini Maglite™, which retails for about $17. It provides a bright, sharp, efficient light using two 'AA' batteries and a halogen bulb, and it can be used for area lighting.

You may also have seen flashlights with a built-in generator operated by repeatedly squeezing a lever. Durability and fatigue may both be issues if you need to rely on these for more than a few minutes, but batteries won't be.

❊ Batteries

The December 1997 issue of *Consumer Reports* has a section on single-use vs rechargeable batteries. One of the conclusions is that cheap, no-brand-name alkaline batteries can be very cost-effective.

Remember that if the power goes out, rechargeable batteries and rechargeable flashlights need to be fully charged to be useful, and that once discharged they'll be useless until the power comes back on.

When considering batteries for flashlights, bear in mind that lithium batteries have a longer shelf life than alkaline batteries, although they are far more expensive. Check 'best-before' dates. And remember that a flashlight using a high-tech krypton or halogen bulb will provide more light and double or triple the life of the battery.

Remember also that expended batteries should be treated as toxic waste because of their acid and heavy-metal components. Keep a glass jar handy for dead batteries, and save them for a 'toxic round-up' in your community or take them to a toxic waste disposal centre rather than putting them in your household garbage.

❈ Failure lights
Some rechargeable battery lights are designed to be left plugged in and to come on as soon as power goes out. Most will then stay on for several hours. For example, Canadian Tire has the Le Lux™ model priced at about $46.

FLAME LIGHTS
❈ Warning
As we noted earlier, burning anything—including candles or flame lamps—produces carbon monoxide and increases your risk of fire, so these lights should only be used in well-ventilated areas. Place them well away from traffic and all objects, especially flammable objects, on steady surfaces that will help protect them from tipping. Extinguish lights before sleeping.

Please refer to notes on storage and use of fuels under 'Fuel Considerations,' above.

❈ Kerosene and coal-oil lamps
These need frequent attention and adjustment to burn cleanly and brightly, but they throw light over a large area. Some models are available from hardware and outdoor stores for as little as $10.

❄ Aladdin™ lamps

These lamps are much more efficient than standard kerosene lamps.

For example, one low-cost model provides as much light as a 60-watt tungsten bulb, with no pumping or start-up hassles. It burns about twelve hours on one litre of lamp oil.

Their only drawback is that they burn hot and must not come in contact with combustible materials.

If you plan to rely on this sort of lamp, be sure to buy a backup mantle and chimney, and wicks, in advance.

The popular Aladdin™ Genie II™ may become hard to find, but we recommend it for situations where you have the space and ventilation to use it safely. Price: about $80.

❄ Candle lanterns

Open candles can be knocked over, brushed against, or set down in a dangerous spot and forgotten, and they are especially dangerous if left burning in a sleeping household.

But those general comments do not apply to lantern models that enclose the flame in an aluminum and glass 'cage,' or lantern. Candle lanterns, in general, are far safer if used as directed.

Most lanterns can either be hung up or stood upon a stable surface. They are available in lightweight models used by campers and heavier models designed for homes.

Prices: The Aurora™ candle lantern, for example, ranges from $6 to $17. A three-candle Candlelier™ is also on the market, priced at about $40.

❄ White gas lanterns

We do not recommend the use of liquid-fuel white gas lanterns inside a living space. The lanterns provide bright light, but they use highly volatile white gas supplied from a pressurized tank.

Under certain conditions, these lamps can produce dangerous amounts of carbon monoxide. Refilling and lighting can be tricky, especially if the lamp is hot, leading to considerable risk of fire. Never store or pour white gas inside your dwelling.

Lanterns, extra mantles and wicks, and white gas supplies are available from sport stores and some hardware stores. Price: about $70.

❄ Propane lanterns

These lanterns can produce carbon monoxide, but in much smaller quantities than you risk with propane heaters. If you use a propane lantern, make sure that you have adequate ventilation, keeping in mind that any flame is a source of indoor pollution.

Don't use *any* flame lantern in your survival space.

Propane is a much safer fuel than white gas. These lanterns are comparatively easy to light, and you never need to pour the fuel.

Several models are available including:

The Coleman™ 2-Mantle Propane Lantern. One canister of propane lasts up to 18 hours on low, 7 hours on high. Price: about $48.

The Coleman™ Electronic Ignition Propane Lantern. This lantern has fully adjustable light levels and needs no matches. It burns up to 18 hours on low, 7 hours on high. Price: about $57.

WATER AND FOOD

WATER

If the power goes off, you may or may not still have running water.

If you rely on a municipal water supply, it relies in turn on electricity for purification and pumping. Many water utilities do have backup systems, because they have experience with blackouts or brown-outs. But, in most cases, their backups have not been designed for longer-term use, and they depend upon supports such as a fuel supply that may be disrupted.

If you use well water or a cistern that is periodically filled by a delivery truck, then you have different problems. You may have one or more electric pumps in your system. Your delivery service may be disrupted.

Regardless of where your water comes from, if your home cools you may have to shut off your system and drain the pipes to avoid freezing damage.

CRITICAL NEED

While we often focus on food, the most important element for your body's survival is water. Most people would be in very serious trouble after three days without water.

Bare survival requires about two litres per person per day. Children, nursing mothers, and the ill need even more.

Since you also need water for basic hygiene and food preparation, we recommend that you store at least *four litres per person per day*.

If you double that to eight litres per person per day for all uses, you will need to store 56 litres per person for a week without your normal water supply. In a household of four, for example, that's a lot of water.

When you calculate how much water you plan to store, don't forget to factor in pets and other animals that count on you.

SAFETY CONSIDERATIONS

Stored water may be contaminated by its containers or by things that get into the containers, so if you're storing water, make sure the containers are clean and well sealed.

Never store water in containers that have held toxic substances, such as gasoline or household chemicals, no matter how well you wash them out.

Power failures may disrupt water treatment as well as supplies, and if this happens you should listen to your radio before assuming that your tapwater is safe to drink without purification.

Water can be stored relatively safely for six months. If it becomes cloudy or murky, either discard it or treat it as described below. Many people store tap water in reusable plastic containers such as soft-drink bottles, but note that some plastic milk containers are biodegradable and may leak over time, especially if the contents freeze.

Avoid glass containers because of the chance of breakage.

Mark the containers with the date they are filled, and discard after six months. Remember that impure water can be used for plants, the toilet, and so on—don't waste it.

If you prefer to buy bottled water, remember that your favourite supplier may have the same problems you do in the event of Y2K disruptions. But until then, bulk containers of water can be purchased from supermarkets and water vendors for about $7 for 20 litres, plus bottle deposit.

Remember that water weighs about a kilogram per litre. Whatever size containers you choose, be sure they can be easily lifted by more than one member of the family. You don't want to spill your supply or add a back injury to your worries.

STORAGE

The most important thing to keep in mind is the potential for freezing.

❋ Will the only non-freezing area in the house be your warm space?

You will need enough room to store your water containers, or you'll need a second warm space. In this

unoccupied space, a kerosene heater on low should keep
the temperature high enough to keep your water liquid.

✸ If your containers freeze, will they burst? You can lessen this
risk if you don't fill them to the top. Freeze one as a test.

✸ You can use large containers, such as plastic garbage cans,
to store water, but remember that you won't be able to
move them once they are full.

✸ Filling your bathtub adds to your water supply—your tub
may hold fifty or more litres for hygiene use, but not for
cooking or drinking unless it is purified.

PURIFICATION

✸ The safest way to purify water is to bring it to a rolling boil
for five to ten minutes.

✸ You can instead add four drops of bleach per litre of water
and stir. Regular household bleach (usually a solution of
5.25% sodium hypochlorite) will kill most microbes. Don't
use bleaches with added cleaners, 'colour—safe' bleaches, or
scented bleach.

✸ Iodine-based water treatments commonly sold at camping
outlets are effective, but they're not recommended for prolonged use.

✸ If you decide on a filtration device, choose one that does
not require electricity.

✸ Seal your containers of pure water, label them, and store
them in a cool, dark place.

✸ **Water purifiers**

Carafe water filters are available that remove contaminants such as
organics and lead, but not parasites. All you have to do is pour
water into the two-litre pitcher. Although convenient, they're
expensive for large quantities of water, and they're quite slow.
Two models discussed in the July 1997 issue of *Consumer Reports* are
the Ecowater™ 2500I and the Brita™ Standard 35507.

Faucet-mounted water filters improve water quality when the
water is running but the power is out. They remove organics and
lead, but they tend to clog up with serious pollutants. Faucet models
can be bypassed when you don't need pure water.

Portable water purifiers have come of age. Current models give a good flow (about one litre per minute) and make the 'safest' water. We recommend two models well tested by trekkers in third-world conditions:

* The PUR™ Voyager Purifier removes micro-organisms, bacteria, and viruses. Price: about $86. A filter cartridge cleans about 375 litres, and replacements cost approximately $53.
* The Sweetwater™ Guardian Microfilter is another reliable filtration system. Price: about $56. A filter cartridge cleans about 750 litres, and replacements cost about $37.

WATER FOR OTHER PURPOSES

Beyond the four litres per person we consider essential, most of us use 100 litres or more per day. We take showers, flush toilets, wash clothes, clean house, and so on. Obviously, in an emergency situation, you can drastically tighten up on how much water you use, to make your stored supply last.

You may have to pile up dirty clothes, wash dishes in a small basin, and maintain a dry bathroom for a while—anything to cut water consumption.

See the 'Sanitation' chapter below for specific advice.

LESS-OBVIOUS WATER SOURCES

If you get caught without an adequate water supply, there may be some at hand:

* Your hot-water tank. To use the water in your hot-water tank, be sure the electricity is off, and then open the drain at the bottom of the tank. Start the water flowing by turning off the tank's water-intake valve and turning on a hot-water faucet somewhere in the house. Do not turn on the electricity when the tank is empty, or you risk damage to your tank.
* Your pipes. If you haven't already drained them to prevent freezing, you may be able to collect several litres by

draining your pipes when you need the water. See the
'When to Worry About the Plumbing' section.

* Toilet reservoir tanks. Make sure you purify this water
before using it.

* Water beds. If you have a water bed, you have hundreds of
litres of stored water, as well as having something else to
drain if your house freezes. This water may contain toxic
chemicals, so don't use it for drinking, washing yourself, or
food preparation.

* Snow. Melt it first! You risk hypothermia by eating large
amounts of snow for water, especially if you're already
chilled. In most areas, you should purify the water that you
get from melting snow, and in some areas, because of air-
borne pollution, snow may contain toxins that you can't
remove with filters, chemicals, or by boiling.

FOOD

Every gram of our food—except what we grow, hunt, or gather
ourselves—comes to us at the end of a very long and complex
transportation chain that often extends halfway around the world.
Whether you live on Vancouver Island, in La Ronge, Saskatchewan,
in central Ottawa, or on the coast of Newfoundland, you depend
on other people and finely tuned electronic systems to bring you
your daily bread, let alone bananas.

If we do suffer Y2K disruptions, Canadians may discover they're
living in a relatively well-functioning country that nonetheless must
cope with interrupted supplies from outside, or we may have a
more serious domestic situation where power, heat, water, communications, and transportation are disrupted within Canada.

We recommend that you plan for both possibilities by storing a
month's supply of good, nutritious food that will be easy to store
and cook even if the power's out. This supply can help you weather
either a serious short-term disruption, or a longer-term period
of 'on-again, off-again' supplies.

You probably already have enough food in your home to keep
you going for a week. Why not add enough carefully chosen items
to prepare your household for a month of disrupted supply?

BACK TO BASICS

Our recommendations come out of experience with dietary requirements for expeditions to remote and cold environments. In those circumstances, you make tradeoffs between nutritious and high-calorific foods, and ease of preparation versus expense, and catering for individual tastes and needs.

In preparation for possible Y2K difficulties, we encourage you to think about how you would deal with the practical difficulties of cooking in the open air on an extended camping trip. We assume that most of our readers will not have access to a wood kitchen range or generator for powering small appliances.

✳ Starting point: your fridge

Remember, if your heat and power go off, the insulation in your fridge will keep things from freezing for quite a while in a cold house. At some point, though, everything in your fridge that can freeze will freeze. Use the perishables and products susceptible to freezing first. Remember to transfer glass jars and canned goods to your warm storage space when necessary, or to discard them in a way that won't cause a mess or a danger when they freeze.

✳ Advantage Canada!

In many parts of Canada, winter provides a reliable freezer in the great outdoors. Unless you're living in an apartment without a balcony, you can probably take advantage of this fact.

Anything in a supermarket freezer—such as vegetables, juices, ice cream—and any other freezable food—such as meat, bread, milk in cartons, and pre-cooked meals—can be stored outside, provided, of course, that it's not available to the world! No sense in feeding the neighbourhood's pets and wildlife.

✳ Foods that won't survive freezing

Carrots, potatoes, and other root vegetables store well as long as the temperature does not drop below freezing, but most vegetables turn positively ugly if frost-nipped.

Even some foods that have been processed for long-term storage have problems if the temperature where they're stored drops much below zero: for example, cans may split and become contaminated and useless, and glass jars may break.

As described above in the 'Heat' section, planning a warm storage space for food and water is a good idea. Your aim should be to provide sufficient heat to keep the space a degree or two above freezing, no more.

HOW TO START YOUR FOOD PLAN

❋ Dinners

We suggest that you draw up a menu based around a weekly rotation of foods that suit your needs and preferences.

As a sample menu, you could plan:

- ❋ 3 dinners based on rice
- ❋ 2 dinners based on pasta
- ❋ 2 meals based on potatoes (fresh, frozen, instant, or dried).

Add meat, fish, vegetable substitutes, packaged sauces, seasonings, and vegetables for nutrition, taste, and variety, but do your best to keep it all in one pot.

Be flexible and creative with your basic menu and don't waste any liquids or food scraps that have value in a stew or soup.

❋ Breakfast

This an important meal in cold conditions, just like your mother told you. You must start off the day with adequate fuel. Cooked porridge is a good source of energy, especially when you fortify it with nuts and sunflower seeds and add milk and sugar. Hot porridge is better value than cold granola.

Plan on having a couple of slices of buttered bread with honey or jam, and have at least two cups of a warm drink: tea, coffee, or hot chocolate.

❋ Lunch

This can be as simple as bread with cheese and salami, plus a warm drink. Or better still, have a bowl of soup or instant noodles with added goodies, such as leftover potatoes and vegetables. Don't deny yourself a cookie or two.

❄ Fat and cold

Mountaineers and polar explorers know that the body demands a much higher fat intake in extremely cold conditions. If you're living in a cold house, you might need to take a winter vacation from cholesterol concerns in order to keep your body fueled. If you have a specific concern, make sure you get your doctor's advice.

❄ How much food to store

We deliberately haven't advised on quantities—there are just too many variables and individual factors to take into account. You know how much food your house goes through in a month better than we do.

See the checklist section for a sample shopping list.

FOOD AND COOKING TIPS

❄ Store foods that you normally eat and enjoy, but concentrate on the ones easiest to cook and to store for extended periods.

❄ You're the expert on quantities for your own household. If necessary, monitor what you eat over a few weeks to calculate quantities.

❄ Don't forget to account for special diets, allergies, and pets in your planning.

❄ Fill your freezer and cooler with packaged frozen foods. You can store them outside if you live in an area where you can count on freezing temperatures in mid-winter.

❄ Avoid cans and bottles—they may burst if they freeze.

❄ Store food as much as possible in rodent-proof, waterproof containers, especially if you plan to store it outside. Start collecting food-grade plastic buckets that seal really well, and these can be your storage containers for staples and other dried goods.

❄ Put lots of staples aside. It's easy and inexpensive to build your supply of rice, pasta, flour, oatmeal, dried beans, powdered milk, and other dried foods.

❄ Take advantage of sales, case-lots, and other economies over the course of 1999 to reduce the impact of buying for an extra month. You could get together with neighbours or friends to make the process more enjoyable.

* Buy specialty items early. Some suppliers are already having difficulty filling orders.
* Rotate your supplies. When you've stored extra food, use the oldest stock first, and replace it with fresh supplies as you go along through 1999.
* Look for instant meals—the 'just add hot water' kind. You can also plan 'boil-in-a-bag' meals of your own: they can be delicious, simple to cook, and quick to clean up even in a frozen kitchen.
* Slip a few goodies into your cupboard to ease the monotony if you end up relying on your pantry for a month. Treats are important morale boosters.
* Soups, stews, chilies, noodles, and rich sauces warm the body and the soul.
* Juices and milk in 'tetra packs' will survive freezing and can be stored on the shelf.
* You can't store fresh fruits and vegetables for long, especially in freezing conditions, so remember to emphasize frozen and dried fruit and vegetables on your shopping lists. If you run through those supplies, lentils and alfalfa can be sprouted in jars to add fresh vegetables to your diet.
* You may consider purchasing a dehydrator to make your own dried fruit and meat jerky. Store the results in zip-lock bags labelled with the contents and the date stored.
* Eating off paper or plastic and using paper towels can reduce your use of water for dishes, but this only works for the short term.
* Aluminum foil is really handy for cooking and storage.
* Have a good supply of matches stored in a waterproof container.
* Make sure you have a manual can-opener.
* A pressure cooker saves fuel and cooking time.
* Fill your insulated mugs and Thermos™ flasks when the pot is hot.

HEALTH

CARBON MONOXIDE POISONING

Carbon monoxide (CO) does not advertise its presence—this highly poisonous gas has no odour, taste, or colour. It is formed during the incomplete combustion of any carbon fuel, including wood, candles, natural gas, coal, fuel oil, kerosene, and propane.

Burning fuels without adequate ventilation produces more CO. In today's almost airtight homes, any device that burns without a fresh-air supply from outside the house and a proper vent or chimney to the outdoors will produce CO that will remain trapped inside.

CO molecules attach to red blood cells even more readily than does oxygen, preventing oxygen from reaching your body tissues, including your brain and vital organs. Your body becomes starved for oxygen. In high concentration, CO is an immediate problem, but it can also build up slowly in the blood over long periods because of its strong affinity for your red blood cells.

At 400 ppm, a healthy adult risks death with three hours' exposure, and 100 ppm can cause death with eight hours' exposure—while you sleep, for example.

If anyone exhibits these flu-like symptoms, you should suspect carbon monoxide poisoning *first*:

Low concentrations of CO:
* slight headache or trouble breathing during moderate activity
* nausea
* vomiting

Higher concentrations of CO:
* severe headache
* dizziness
* disorientation
* rapid heart rate
* fainting during exertion
* drowsiness

At the first sign of these symptoms, get the affected person outside into the fresh air. If the person can't be moved, open doors and windows to allow fresh air to circulate.

Then call 911 or the nearest Poison Control Centre immediately.

If your CO alarm goes off, even if you don't notice any of the above symptoms, trust the device and go outside. Don't go back inside unless an expert confirms it is safe to do so.

HYPOTHERMIA

Hypothermia posed a serious public health risk for people in their own homes during the 1998 ice storm—at one point a check of thirty-eight residents thought to be at risk found four suffering from severe hypothermia, and fourteen others who needed to be strongly convinced to leave their residences.

Cold, moisture, and wind cause hypothermia, especially in combination.

Beware of these symptoms:

* As the body temperature drops, fatigue and shivering can set in. Slurred speech and disorientation may follow. If the body temperature drops too far, the victim can lose consciousness.

If you notice any of these symptoms:

* Bring the victim to the warmest place possible and prevent further heat loss. Direct body contact with the victim in a warm sleeping bag or under a duvet may be ideal.

✳ If the victim can swallow, administer warm, non-alcoholic drinks.

✳ Get medical help as soon as possible.

Parents should be particularly alert to the symptoms of hypothermia in small children, especially those still in diapers. It may be difficult to detect early symptoms such as slurred speech and disorientation in the very young, or in someone unable to speak.

Ensure that your children and other vulnerable people are kept well bundled if the temperature in your house or warm space drops below normal levels. Check for wet diapers regularly—the heat loss from damp clothes is a prescription for problems.

While there's no reason to prevent normal play and activity, ensure that, once indoors, children's clothes are dry and adequate for the conditions.

GENERAL HEALTH CONSIDERATIONS

Nothing in this section is intended to be a substitute for sound medical advice from a physician.

If you have an existing condition, then you should talk to your physician about dealing with it in times when health-care services may be disrupted or overstressed. You may need a contingency or backup plan.

Do your best to anticipate your needs, prepare yourself to be as independent as possible, and check with your doctor or other supports about their readiness and anticipated availability early in the New Year.

PREPARE WHAT YOU CAN

While our health-care system is preparing to the best of its ability, there is still the possibility that high demand or disrupted infrastructure could make emergency services or hospitalization difficult in the event of a health emergency. Get yourself as healthy as you can, and make whatever advance preparations you can, so that you minimize your reliance on the health-care system during a period when it may not be functioning smoothly.

MEDICAL AND DENTAL CARE

Try to schedule examinations, tests, checks, teeth cleaning, and other procedures in the last three months of 1999. If you have any medical or dental problems you've been putting off, take care of them now.

Check with your doctor to ensure that he or she is well aware of the Y2K problem. Ask if the office has paper backup of your records.

If you have elective surgery scheduled just before or just after the turn of the century, discuss the possibility of postponement with your doctor. Some hospitals may themselves decide to cancel elective surgeries around the end of 1999 and resume early in 2000 when they can see how their systems and services have been affected.

EMERGENCIES

Y2K problems may make it difficult to contact emergency services or transport someone to a hospital.

If you can't contact emergency services you may need to transport someone yourself. Ensure that you know the location of the nearest emergency facility and the best way to get there. Remember that you may face longer waiting times when the system is under stress.

MEDICAL CONDITIONS

The frail, the elderly, and people with particular medical problems or disabilities may need to make special plans for their safety in the event that emergency services are disrupted.

If you know of someone who might be at risk without power, heat, or telephone, ask neighbours and friends to watch the situation and offer assistance if necessary.

Make backup plans for members of your household who require regular treatments such as dialysis. Talk with your doctor or clinic.

Modern battery-operated wheel chairs are good for around forty-five miles between charges and should survive a short-term blackout. If any power interruption appears to be lasting longer than your battery power, you may need to arrange to get to an emergency centre or shelter to recharge. If you rely on some other electrical or battery-operated device, such as a ventilator, arrange in advance for a backup unit or for an alternative recharging system in case of power failure.

PHARMACIES AND MEDICATION

Pharmacies rely heavily on computers to store your prescriptions, medication records, and insurance information. Y2K-related errors could make it difficult for you to get needed medications. If you can pre-fill prescriptions in the last few months of 1999, then do so. If you do fill prescriptions in early 2000, check all the information very carefully and question any changes, in case the system has made an error. Since automatic insurance coverage may not be functioning, you'd better bring along some cash (see the 'Finances' section below).

❋ **Further reading**
  www.redcross.ca/emergency/y2k/y2k.htm

ASSESS YOUR SITUATION

What will you need? This section includes a basic first-aid kit, and it will help you decide what you may require in addition to your kit.

  ❋ List all medications that you use: prescription and non-prescription plus related supplies.
  ❋ List all devices you use at home or at a medical facility (for example, dialysis machine, pacemaker, glucose testing equipment, inhalers, respirators, and so on).
  ❋ Do you use medical devices at home that require electrical power? If so, you need to investigate an alternative power supply in case of a power failure.

Discuss the following issues with your doctor if necessary:

  ❋ Order enough medication and other medical supplies in advance and store them safely.
  ❋ Can you manage your medical condition in an emergency? Do you need to purchase additional equipment or get training?
  ❋ Check with the manufacturers of the devices you rely on for a statement of Y2K compliance. Manufacturers should

share compliance information for the devices that they sell; if they won't, register your complaints loudly and in writing and talk with your doctor.

* Find out from your doctor what you can do if a necessary device doesn't work properly or fails.
* Do you need to have spare parts available for any devices? If so, buy them soon.
* Do you know how to make basic repairs if necessary?
* Discuss these issues with members of your household so that you have some backup. Get everyone involved.
* Test any emergency plans that you develop.

Training in first aid, CPR, and other emergency skills will leave you feeling far more secure if health services become stressed. Consider taking what courses you can, then look to your neighbours, friends, family, and community so that you know where to find the closest trained people in your neighbourhood.

EXERCISE
Being without heat doesn't mean that you have to become a couch potato. Try to maintain your normal level of activity and exercise, even though a sponge bath in a cold house may not be everyone's perfect end to a work-out.

BABIES AND SMALL CHILDREN
Babies need special care in a winter emergency.

Make sure that you have a good supply of baby food or formula, supplies, and diapers. Moist towelettes can make life much easier if bathing becomes difficult.

If you do bring any flame devices into the home, you have to increase your vigilance when any children are around. Keep fire out of reach and ensure that adult supervision is available at all times.

While anyone is susceptible to carbon monoxide poisoning, unborn babies, small children, and the elderly are more susceptible than healthy adults. As recommended above, purchase a battery-operated CO monitor with a digital read-out of the parts per million (ppm) of CO in the air. This way you can watch any increase in CO levels and take action earlier at lower levels with high-risk people in your home. Fifty ppm is the maximum allowable concentration for healthy adults exposed over an eight-hour time period; longer exposure makes that level unsafe.

And, as described above, hypothermia can be another worrying condition with small children.

BASIC FIRST-AID KIT

Ensure that you have the best basic first-aid kit you can afford. Add to it if you can.

This list is based on Red Cross recommendations. It will cost about $100, plus the cost of whatever non-prescription drugs you include.

Tick off each item as you purchase it.
- ○ Sterile adhesive bandages in assorted sizes
- ○ Hypo-allergenic adhesive tape
- ○ Assorted sizes of safety pins
- ○ Cleansing agent/soap
- ○ Two pairs of latex gloves
- ○ Sunscreen
- ○ 5-cm sterile gauze pads (5)
- ○ 10-cm sterile gauze pads (5)
- ○ Triangular bandages (2)
- ○ 2-inch sterile roller bandages (2-3)
- ○ 3-inch sterile roller bandages (2-3)
- ○ Scissors
- ○ Tweezers
- ○ Needle
- ○ Antiseptic
- ○ Antibiotic ointment
- ○ Thermometer
- ○ Tube of petroleum jelly
- Non-prescription drugs:
- ○ ASA or non-ASA pain reliever
- ○ Anti-diarrhea medication
- ○ Antacid
- ○ Laxative

○ Activated charcoal
○ Syrup of Ipecac (use to induce vomiting if advised by your local poison control centre)

Other items you might want to consider if you have the resources:
○ Butterfly bandages
○ 3-inch elastic bandage
○ Sterile eye patches
○ Sterile cotton balls
○ Cotton swabs
○ Small plastic cup
○ Chemical cold packs
○ Paper cups
○ Space blanket
○ Antiseptic/anesthetic spray
○ Calamine/antihistamine lotion
○ Sterile eye wash
○ Packet of tissues

SANITATION

It's possible that normal garbage collection and related services will be disrupted for a period of time in early 2000. You need to be prepared to deal with accumulations of garbage for two weeks to a month. In a cold climate, this shouldn't be too difficult, but it would be wise to have extra containers with secure lids available. And, as always, try to minimize your garbage.

Body waste—sewage—may pose a problem if you have no heat, and especially you're also without running water. You may want to buy a chemical toilet, or a portable toilet with a large supply of plastic bags, as a backup plan.

Examples:

* ❄ Folding camp toilet. Price: about $22; extra bags about $3.25.
* ❄ Luggable Loo™. Price: about $16.
* ❄ Hassock™ Camp Toilet. Price: about $32.
* ❄ Thetford Porta Potti™. Price for 20-litre model: about $149.

PERSONAL HYGIENE
Stay clean, stay healthy.

Washing your hands after using the toilet is a simple and effective sanitary measure to prevent stomach upsets and more serious problems.

A constantly available basin or bucket of water, treated with a little liquid bleach, together with antiseptic soap, should help everyone remember the basics. Moist towelettes are very useful for adults as well as for babies. A wash-cloth and a basin of warm water can substitute for a bath or shower if necessary.

Personal hygiene supplies are covered in the 'Supplies and Tools' task list.

COMMUNICATIONS

PHONE

Telephone companies are working to ensure that everyone hears a dial tone when they pick up the phone next January, and emergency services and communications for the electrical utilities are at the top of their priorities list. Like everything else involving the Y2K bug, nothing is guaranteed, so we recommend being ready to deal with a world without telephone communications, at least for a short time.

If there are phone service interruptions, you could be cut off from emergency medical, police, and fire services. Cellular phone services might survive for longer than regular phone services, but they may also become unavailable. CBs may be a very useful commodity if dial tones do disappear for a while.

The phone companies will probably request that you don't lift up the phone just after midnight on New Year's Eve to see if you have a dial tone. That in itself could cause problems if done simultaneously in thousands of households. If there are serious Y2K issues across North America, the telephone lines may be overloaded by distress calls and 'are-you-OK' calls among family members. So consider minimizing your calls to family and friends until later in the day next January 1st.

RADIO

Make sure that you have a battery-operated radio, with an ample supply of batteries. Your radio will help keep you informed about the extent of the disruption and how long it is likely to continue. Monitor the news so that you can update your plans accordingly.

Keep in constant touch with your neighbours—someone somewhere may be a ham-radio enthusiast and be able to provide you with more information than you are picking up on the 6 o'clock news. But remember the difference between rumour and information.

TRANSPORTATION

We recommend that you do some thinking in advance about personal and public transportation in the early days of the year 2000. Traffic signals, train and airplane control systems, and ships rely heavily on computers and embedded microprocessors. Walking and bicycling may be the most reliable bets early next year.

ASSESS YOUR SITUATION

Where are you planning to be at midnight, 31 December 1999? Are you going to be at home, at a local party, at some festivities in the city where you live, or at festivities in another town, or another country? Are you planning to fly or cruise somewhere?

If you are planning to be away from home—whether at a distance of minutes, hours, or days—you may want to consider contingency plans for communication and transportation home. Services may not be reliable in the early part of the year 2000. Home is where the heart is. Perhaps block parties would be a wiser choice this New Year than any celebration that requires travelling longer distances.

You may already have big plans in place and be unlikely to change them, or you may be willing to change your plans as part of your overall Y2K strategy.

We recommend that you base your decision partly on your evaluation of where adequate planning can cope with potential problems.

YOUR EMPTY HOME

Unless you've arranged for a reliable person to check your house regularly, and unless that person is prepared to deal with the electrical and plumbing systems in the event of power problems, we recommend that you prepare your house to sit empty. That may mean draining the pipes and 'evacuating' items that would be damaged by freezing.

Remember that most insurance policies do not cover a dwelling left uninhabited for longer than 24 hours. And some policies may specifically exclude Y2K-related damage. See the 'Home Insurance' section below.

THE PRIVATE CAR

Most car manufacturers are issuing comforting statements that state owners will have few problems with private automobiles. Cars will continue to run, they say, although the micro-chip problem is a little more complicated than 'my car doesn't care what day of the week it is.' As a contingency plan, we recommend that you keep more than one option open if you absolutely need to travel early in the year 2000.

Remember also that you'll be travelling in mid-winter, perhaps in blizzard conditions, and both your cellular phone and roadside emergency phones may not be working.

Make sure your car is well winterized this year; it may need to start without the benefit of the block heater. We also recommend that you do what smart drivers already do in Canada: carry sufficient warm clothing, blankets, candles, and snacks to be able to cope with a breakdown situation. You'd also be wise to keep emergency tools and a first-aid kit (see 'Health' section) in your car. Include:

* booster cables
* flashlight
* blanket
* fire extinguisher
* chocolate bars, granola bars, cold bars, or some other form of nourishment
* shovel
* flares
* a spare tire in good condition, properly inflated.

Keep your gas tank topped up; this is especially prudent in the early days and weeks of 2000 when gasoline supplies may be restricted or the gas stations may be closed if there is no power.

PUBLIC TRANSPORT

Buses, streetcars, and trains may have problems.

And don't forget traffic control systems. They are an obvious hazard, so most cities should have them all assessed, fixed, and tested again by the end of 1999. Watch the news as 1999 wears on.

AIRLINES

Planes have a myriad of microprocessors on board, and air-traffic control systems all use networks of interconnected computer hardware and software.

Some airlines say now that they will not fly unless they are certain that all aspects of the system in the countries that they fly over are Y2K compliant.

The Chinese are reported to have taken an interesting motivational approach: they have mandated that their airline officials be in the air to welcome the year 2000.

We recommend that you stay on the ground if at all possible for the period immediately before and after the New Year, until you see how the system is coping.

CRUISES

If you have booked a New Year's cruise, check with the cruise line on the status of their Y2K remediation activities. A cruise ship is like a small town: many internal systems rely on computers and embedded-chip technology, such as:

* the heating, ventilation, and cooling system
* stabilizers
* process-control devices
* navigation systems.

While cruise lines are well aware of the issues, your check of how they've handled compliance issues will be very reassuring.

FINANCES, RECORDS, AND INSURANCE

HOME INSURANCE

The insurance industry is positioning itself for survival after Y2K. Every company is approaching the Y2K problem differently, and they're all aware that the potential for claims ranges from minor to massive. Some are now attaching exclusionary endorsements to all policies; others underwrite each class of insurance individually.

To add to the confusion, some losses that occur as a result of a failure caused by the Y2K bug may be covered even if the policy contains a Y2K exclusion!

We urge you to contact your own home insurance agent for clarification of the company's updated exclusions, if any.

As we write this, it appears that if you have devices or controls with embedded computer chips within your own home—such as the thermostat, home-security system, or fire suppression—it is your responsibility to check their compliance and make any necessary changes. Your home insurance will not pay for the cost of updating or repairing those systems, but it may cover damage to other property resulting from a Y2K failure.

If the problem is caused by a failure *beyond your control* (for example, if your pipes freeze and burst because of a municipal power outage), then your normal coverage may apply. Again, check your policy for 'normal' coverage, and ask your agent for clarification where necessary.

Nearly all policies have a frozen-pipe exclusion that states that you will not be covered if you leave your home for longer than 96 hours during the season when heating is required; you must arrange for a competent person to enter the house daily to check the heat. Your alternative is to turn off the main water valve and drain the entire plumbing system.

In case your house does suffer extensive damage, make sure you have adequate records of your possessions and their values to substantiate a claim in case of loss, and keep your records in a safe place, such as your safety-deposit box.

KEEP RECORDS

Leading up to and immediately after the New Year, keep hardcopies of all bank statements, mortgage papers, car and rent payments. You may need to prove your position.

* Keep good records of your tax processing in the year 2000.
* Keep cancelled cheques, debit slips, and credit slips.
* Get hardcopy receipts for purchases for your records.

STORE IMPORTANT DOCUMENTS

In case your computerized account records become temporarily unavailable because of the Y2K bug, you will want to safely store paper print-outs and recent transaction slips, plus your other important documents, before the New Year.

Include:

* mortgage records and proofs of ownership, such as deeds,
* bank-account numbers and current information,
* loan-payment records,
* credit-card and debit-card information, including transaction receipts since your last statement of 1999,
* proof of rent payment,
* insurance agreements,
* birth and marriage certificates,
* divorce papers and agreements,
* RRSP statements,
* Social Insurance, Employment Insurance, and Canada Pension information,
* tax returns,

* contracts,
* wills,
* passports, and
* any other official documents that you may need.

FINANCIAL SYSTEMS

Our focus in *Weathering Y2K in Canada* is personal readiness. Finance issues may become very complicated if the Y2K bug seriously affects national and international economies, but our recommendations here relate only to the short-term plans for limited interruptions in supplies and services. We leave considerations of longer-term effects to other texts, which you may want to explore. We provide references for further investigation at the end of this section.

❅ Cash

Cash often works best in a crisis, so as difficult as it might be, we recommend that you set aside some cash as a Y2K backup. Without causing yourself severe hardship, try to set aside enough for two weeks. It is possible that credit cards, debit cards, and personal cheques may not be accepted early in the New Year.

You must make your own decision about how much cash to set aside. The right amount will differ for each individual or family. Most people know their cash flow for a month, since for most people it's approximately what they earn!

Don't forget that you may be able to offset your need for cash by stocking up ahead of time. Prices of some items may increase if there are significant problems with supply networks, so this may also save you money.

Storing cash at home is a dangerous business. While some Y2K guides suggest spreading your cash around the home in all manner of 'safe places,' we believe it would be an invitation to crime to suggest that everyone keep hundreds or thousands of dollars where they live.

We recommend that you take the chance that the banks are not going to be completely closed for any great length of time, and put your cash in your safety-deposit box. Many vaults have computer-controlled access, but most banks control access to the boxes themselves with old-fashioned signatures, and the boxes open with keys. If you deal with more than one bank, you could arrange for a safety-deposit box in each for a short period of time as a way to hedge your bets.

Don't keep anything at home that you can't afford to lose.

Finally, don't forget: when it comes to the Y2K bug, Canadian and U.S. banks are probably the best-prepared institutions in the world. If they're closed, the odds are that you won't have many places to spend your money anyway.

✳ Your income

Since most accounting systems are automated, expect some delays or problems with pay cheques, government benefits, and other income. Having your pay cheque delivered rather than using automatic deposit may not help if the system has trouble generating the cheque in the first place.

If you tend to live from month to month, or from day to day, preparing in advance will make it easier to get through periods when your income may be interrupted. Do your best to set aside some savings, or arrange some formal or informal credit in advance.

✳ Can you pay ahead of schedule?

If missing a crucial payment could cost you dearly, you may want to discuss the issue with your bank manager or creditor to negotiate alternative strategies if your income is delayed for Y2K reasons. Get any agreements in writing.

If you face penalties for pre-payment or late payment, discuss this with your creditor. Ask if they will waive penalty fees if the problem is Y2K related.

✳ Barter

If you aren't able to put aside as much cash as you would like, look into the barter economy. Barter is common practice in many areas. Read up on barter at your local library and acquire goods or list services you may be able to use as currency should you need to.

✳ Debt load

One often-suggested financial-management strategy is to reduce your debt load so that you're more resistant to Y2K-induced variations in interest rates. Since this isn't a bad strategy even when you're not preparing for Y2K, you may want to consider it. If you do have the opportunity to reduce your debt load, now might be a good time to do that.

Know your limits, and only spend what you can manage, but on the other hand, don't neglect Y2K preparedness just to reduce debt. You may need to spend some money to be ready to weather Y2K.

❆ Investments

Investment strategy is beyond the scope of this guide. Your own financial situation and strategy will determine how much research and effort you put into this problem.

If you rely on financial advisors, check whether they're aware of the possible consequences of the Y2K problem.

Beyond that, our only advice is to make decisions that you're comfortable with, whether your personal approach stresses risk or security.

❆ Precious metals

These commodities have in the past provided refuge for investors in times of economic hardship, but there's no specific evidence that we know of to suggest that the Y2K bug will create conditions favourable to owners of gold and silver.

As with other investment issues, you and your advisors must make up your own minds on where best to hold your money as the year 2000 approaches.

❆ Further reading

References for further investigation:
http://www.Yardeni.com/
http://www.Year2000.com/

Y2K-SENSITIVE DEVICES

Manufacturers indicate that very few household devices are non-compliant, but you don't want the exception to be yours.

VCRs, 'smart' coffee machines, and so on will only cause a nuisance if they fail, so they can safely be dealt with after the year 2000.

But you should check the compliance of security systems, programmable thermostats, electronic locks, sprinkler systems, and of course your personal computer if you have one.

ELECTRONIC LOCKS

If you have electronic locks or a security system, make sure you know what the default setting is before the power goes out. Some set to 'open' for fire-hazard reasons, and this adds an extra security concern if your house will be unoccupied when the New Year arrives. Check what will happen to yours.

PERSONAL COMPUTERS

Again, it goes beyond the scope of this book to provide detail on how to manage your PC, except to say that you need to investigate the hardware, the operating system, the packaged software, and any custom applications that you may have built yourself.

If you have a newer Macintosh, you can ignore the hardware and operating system since they are compliant, and focus on your applications.

PC Novice magazine's 'Guide to Y2K' is a 144-page reference manual on how to identify and fix Year 2000 problems. This guide, and others like it, should be on magazine racks all year. You can now buy many programs to check the compliance of your PC, and others to fix problems, so it's probably best to check the computer magazines for recommended products and online resources before you buy.

TOWN AND COUNTRY

HIGH RISES AND APARTMENTS

The tenants or unit owners in an apartment block or high rise form a special class of neighbourhood. If you live in an apartment, you can accomplish much more as a group than as individuals.

Form a Y2K action group with the other owners or tenants and involve your landlord so that you can make assessments and decisions about your building. These should include:

* What will be allowed with regard to heaters in individual apartments? Portable combustion heaters may be illegal, or they may void the insurance in your building.
* Remember our strong recommendation that everyone, wherever they live, can create a survival space that doesn't require supplemental heating.
* Can you create a communal warm space for everyone's use if necessary? You may be able to get the landlord's OK to install backup heat on a semi-permanent basis for the communal warm space.
* Can individual apartments be cold-proofed following the same steps described above for houses, such as draining your plumbing and appliances?
* At what stage will the apartment building be at risk of freeze-up or failure of mechanical services?
* Try to determine in advance when you will need to worry about the pipes. Large buildings will lose heat more slowly than houses, so you may not need to drain your water pipes if the power is not off for too long. Someone will need to monitor the inside temperature and coordinate cold-proofing water pipes and sprinkler systems when necessary.

* Can you arrange collectively for a supply of food and water, and for storage? You may need to help carry water and fuel upstairs for your neighbours on upper floors if the elevators stop working.

URBAN VERSUS RURAL

The Y2K issues you may face will differ depending upon your surroundings.

In cities and towns, the major utilities—power, heat, and water—are almost exclusively provided by central utility companies. In the country, many homes have their own wells or cisterns providing water, and in many districts oil, coal, or propane heaters are more common than natural gas.

Be aware that these more-independent systems come with their own set of problems in the event of a power failure. Wells and cisterns depend on electric pumps which will stop unless you have backup power—220 volt in some cases. Cisterns run dry without regular water deliveries. And some septic systems regularly require electricity to pump out.

Food supplies in rural districts often have to be transported many kilometres to their destinations, and failures in the supply network could create shortages in a matter of days.

On the other hand, rural residents tend to watch out for their neighbours, and they often have good local supplies of food and services. Do your best to get to know the strengths, weaknesses, and state of preparation for the Y2K bug in your area.

NEIGHBOURHOOD AND COMMUNITY

I have no energy to prepare for Y2K in isolation but it is effortless and exciting when I do it in community.

JOHN STEINER,
at Boulder Y2K Conference, August 1998, quoted in *Communities*, issue 101, page 43.

❋ Independence and self-sufficiency

Analyze your personal situation and determine how long you think you can be self-sufficient. You may be in a position to offer a refuge to those around you, but you should also know where you can turn for help if your resources run out.

That 'somewhere' could be your extended family, your friends, your neighbours, or your community.

Y2K problems will be most easily solved if we help each other prepare, and then help each other if the storm hits. The concept we recommend is that if times do get hard, you won't need to protect your 'assets' from unprepared neighbours if they know that there is an 'open door' and a hand extended in friendship.

As you go through your planning process, involve your neighbours. Avoid isolation, share ideas, and you may even be able to share solutions, such as cooperating to buy bulk foods.

Perhaps you'll agree that you can share one warm space among several households. Share skills and expenses while you prepare.

Remember that some people, such as the disabled, may need the assistance of neighbours and friends to make basic preparations.

We anticipate that by mid-1999 many communities will have preparedness plans in development. Check with your community or emergency preparedness office to discover what is going on in your neighbourhood.

FINAL PREPARATIONS

At this point you may be getting concerned about the number of items that you have on your task list. Remember that once you have your whole plan organized, you will be able to tackle these jobs very efficiently, probably in a few weekends. Just do what you can as soon as you can.

Again, we recommend a 'stay-at-home' plan. The more people prepare to take care of themselves in their own homes, the less stress there will be on federal, provincial, and community resources that might be needed to help people who aren't able to fend for themselves.

Do what you can in the months leading up to the end of the year, and take final stock of your situation.

* **Power, heat, light, and cooking.** In the last weeks, it's too late to think about installing a generator or ordering equipment. If you purchased a backup heater earlier in the year, fuel it, test it, and have it ready to go. If possible, try a test run as soon as the mercury dips below -8° C. Turn off the main power breaker and see if your backups work.
* **Water.** Fill your bath and all available clean containers. Store the containers in an area that won't lose heat too rapidly. The basement may be a good bet.
* **Food.** Make sure your bulk supplies are stored where they won't be damaged by freezing.
* **Fuel.** Make sure that you have adequate supplies of wood, propane, kerosene, or gasoline stored safely. If you don't buy them early, you may find supplies limited.
* **Winterize your home.** Install extra insulation and weather-stripping where it will save heat and keep out draughts.

Make sure you have some plywood, lumber, hardware supplies, and tools on hand for late emergencies such as a broken window.

❋ **Liquids are at risk.** Double-check your cupboards and basement for liquids, especially toxic liquids, that could burst their containers and cause safety concerns.

❋ **Lighting.** Make sure you have your flashlights, extra batteries, and other backup lighting supplies where they're easily accessible.

❋ **Cooking.** Organize your backup cooking area (preferably outside), ensure that your backup devices are working properly, and make sure that you have enough fuel.

❋ **Medical supplies.** Make sure that your first-aid kit is well stocked and handy to your warm room.

❋ **Cash and papers.** Make sure that you have all your backup paper—documents and money—in a secure place.

❋ **Pets.** Make sure that you're prepared to supply their basic needs.

❋ **Vehicles.** Gas up your car, check the oil and fluid levels, and check its emergency supplies. If you have an electric garage-door opener, make sure you can operate it without power. Otherwise, leave your car outside.

❋ **Radio.** Make sure that you have a reliable battery-powered radio with spare batteries, since it may become your main source of news, forecasts, and official bulletins.

❋ **Detectors and extinguishers.** Make sure that your fire extinguishers and your smoke and carbon monoxide detectors are in place and working properly. Replace the batteries in the detectors.

❋ **Know your home.** Make sure you know how to shut off the electricity and the water to the whole house, and how to drain the plumbing.

❋ **If you're leaving on holiday.** Make sure that your own plans don't expose you to unacceptable risks, and make sure that your home will be properly cared for in your absence, no matter what the Y2K bug brings.

CHECKLISTS

FOOD SHOPPING LIST
You can use this template to make up your shopping list for a month.

Staples
- ⭘ Rice white, brown
- ⭘ Potatoes, fresh, dried, frozen
- ⭘ Beans and lentils
- ⭘ Spaghetti, noodles

Dairy/Non-dairy
- ⭘ Dried milk
- ⭘ UHT milk
- ⭘ Condensed milk in cans
- ⭘ Cheese
- ⭘ Butter
- ⭘ Margarine
- ⭘ Vegetable oil

Breakfast foods
- ⭘ Oatmeal, porridge
- ⭘ Granola
- ⭘ Nuts and sunflower seeds
- ⭘ Pancake mix
- ⭘ Powdered eggs

Sugar/seasonings
- ⭘ Sugar white, brown
- ⭘ Salt, pepper, seasonings

Vegetables
- ⭘ Vegetables, frozen
- ⭘ Vegetables, dried or dehydrated
- ⭘ Onions, fresh
- ⭘ Tomato paste, pasta sauces
- ⭘ Garlic
- ⭘ Green peppers, dried
- ⭘ Chili peppers, dried

Fruit
- ⭘ Fruit and berries, dried or dehydrated
- ⭘ Frozen fruit, commercial or home-cooked
- ⭘ Dates, raisins
- ⭘ Banana chips
- ⭘ Coconut, dried

Meat
- ⭘ Meat, frozen
- ⭘ Chicken, frozen
- ⭘ Ham, sliced
- ⭘ Fish, frozen
- ⭘ Sausage
- ⭘ Bacon

Drinks
- ○ Tea
- ○ Coffee
- ○ Hot chocolate/Ovaltine™
- ○ Fruit juices, frozen
- ○ Fruit drink powder

Miscellaneous
- ○ Flour
- ○ Baking soda
- ○ Baking powder
- ○ Muffin, biscuit mix
- ○ Yeast
- ○ Soups, packaged
- ○ Instant noodles
- ○ Gravy mix

- ○ Pancake syrup in plastic
- ○ Jams in plastic
- ○ Honey in plastic
- ○ Pickles in plastic
- ○ Macaroni and cheese
- ○ Bread, fresh and frozen
- ○ Cookies
- ○ Chocolate and candies
- ○ Vitamins

Baby food
- ○ Frozen or in plastic

Pet food
- ○ Dried pet food

TASK LIST

Check circles to create your instant plan. Check boxes to indicate completed tasks.

FIRST STEPS
- ○□ Decide how many people you are planning for.
- ○□ Decide how many animals you are planning for.
- ○□ Consider arrangements for those with special needs and list extra tasks.
- ○□ Plan your New Year's Eve strategy.
- ○□ Find out where your immediate family will be.
- ○□ Plan alternative ways home if necessary.
- ○□ Check the travel advisories.
- ○□ Arrange to stay where the party is.
- ○□ Arrange for someone to check your house in your absence.

IF YOU LOSE POWER
Survival space
- ○□ Decide on the location for your survival space.
- ○□ Determine what materials you need to create your survival space.
- ○□ Beg, borrow, or buy the extra materials that you need.
- ○□ Decide where to store these materials in December 1999 so that they are readily available.

○□ Test build your survival space.

○□ Upgrade the basic insulation of your dwelling, especially in this space.

○□ Make sure you have sleeping bags.

Warm space

○□ Decide on the extent and location of your warm space.

○□ Purchase a battery-powered, digital read-out carbon monoxide detector.

○□ Purchase a battery-powered smoke detector.

Warm clothing

○□ Determine what warm clothing you have available.

○□ Purchase extra clothing if necessary.

Safeguarding your dwelling

(If any of these tasks are beyond your capabilities, arrange for someone to come in and help you with them.)

○□ Purchase adequate flashlights and batteries.

○□ Store flashlights in an easy place to find them when it's dark.

○□ Purchase and install emergency lights.

○□ Decide what equipment and appliances you will unplug when there is a loss of power.

○□ Find out where gas and water shut-off valves are for your dwelling.

○□ Obtain and place suitable tools near shut-off valves.

○□ Learn how to shut off and drain your water pipes.

○□ Learn how to drain any appliances that use water, including hot-tubs.

○□ Purchase and install an indoor thermometer.

○□ Learn how to drain your water tank.

○□ Purchase four litres of non-toxic plumber's antifreeze.

○□ Build your plumbing tool-kit.

HEAT, COOKING, AND LIGHT

Warnings

○□ Purchase battery-powered, digital read-out carbon monoxide detector (we know we're repeating ourselves, but we really want you to do this).

○□ Be sure you have at least one ABC fire extinguisher.

- ○□ Be sure you have at least one battery-operated smoke detector.
- ○□ Purchase extra batteries for above.

HEAT

- ○□ Decide on your alternative heat sources.
- ○□ Upgrade the basic insulation of your dwelling.
- ○□ Check and upgrade weather-stripping and seals on doors and windows.

Research your backup heat options

- ○□ Research the rules and regulations regarding heating devices and fuel storage for your dwelling.

PORTABLE HEATERS

- ○□ Purchase kerosene or other portable heater.
- ○□ Determine your needs and purchase sufficient fuel (do not store in your dwelling).

WALL-MOUNTED HEATERS

- ○□ Purchase wall-mounted propane or natural gas heater.
- ○□ Determine your needs and purchase sufficient propane (ensure that you can store this safely and legally).

GENERATORS

- ○□ Decide on size and type of generator.
- ○□ Purchase generator.
- ○□ Purchase heavy-gauge extension cords.
- ○□ Hire a professional electrician to do any necessary re-wiring.
- ○□ Purchase appropriate gasoline containers.
- ○□ Purchase sufficient gasoline.

WOOD STOVES

- ○□ Decide on model, type, and location.
- ○□ Purchase wood stove.
- ○□ Hire a professional to install your wood stove.
- ○□ Decide on dry, accessible location for your wood-pile.
- ○□ Determine your needs and purchase sufficient wood (the earlier the better, so your wood can dry).

WOOD AND PROPANE FIREPLACE INSERTS

- ○□ Hire professional chimney sweep.
- ○□ Purchase suitable fireplace insert.

○ □ Have the insert
professionally installed.
○ □ Decide on dry, accessible
location for your
wood-pile.
○ □ Determine your needs and
purchase sufficient wood
or propane.

FIREPLACE
○ □ Hire professional chimney
sweep.
○ □ Purchase extra wood.

SOLAR POWER
○ □ Research and plan a
solar-power strategy.
○ □ Implement your solar
strategy.

COOKING
○ □ Determine your preferred
cooking alternatives.

**Outdoor barbecues and
camp stoves**
○ □ Purchase propane bar-
becue.
○ □ Purchase camp stove.
○ □ Clean and upgrade existing
barbecue.
○ □ Clean and upgrade existing
camp stove.
○ □ Determine your needs and
purchase sufficient fuel.

Portable stoves
○ □ Research and purchase a
suitable stove.

○ □ Determine your needs and
purchase sufficient fuel.

LIGHT
○ □ Determine your light
requirements.
○ □ Purchase cyalume lights.
○ □ Purchase flashlights.
○ □ Purchase adequate bat-
teries.
○ □ Purchase emergency
failure lights.
○ □ Purchase kerosene lamp
and adequate fuel.
○ □ Purchase Aladdin™ lamp
and adequate fuel.
○ □ Purchase candle lantern
and sufficient candles.
○ □ Purchase other preferred
light source.

WATER AND FOOD
○ □ Decide how many you are
storing water for.
○ □ Decide whether you will
use tap water or purchased
water.
○ □ Decide where you will store
your drinking water.
○ □ Plan to store at least four
litres per person per day
for drinking and food
preparation for the length
of your planning period.
○ □ Gather sufficient suitable
containers.
○ □ Purchase water purifier.

○□ Gather or purchase containers for non-drinking water.

○□ Purchase a chlorine or iodine source for treating your water if necessary.

○□ Purchase heating device for storage warm space.

○□ Purchase sufficient fuel for above device.

FOOD

○□ Discuss with your family what types of food you will store.

○□ Decide on your planning period for food.

○□ Decide whether you'll use frozen foods.

○□ Plan approximate menus.

○□ Calculate quantities of food.

○□ Purchase or gather air-tight containers.

○□ Organize your food-storage space.

○□ Build your shopping lists and spread the purchases over a period of time before October 1999.

HEALTH

○□ Arrange for routine health and dental checks in the last three months of 1999.

○□ Learn and document all local emergency contact information.

○□ Post this information where all family members can see it.

○□ Determine your emergency medical response plan.

○□ Train your family members in this plan.

○□ Determine the plans for any special medical conditions in your household.

○□ Check any medical device issues with your physician.

○□ Develop a contingency plan for at-home medical devices.

○□ Practice plans if necessary.

○□ List any prescription medications that you need.

○□ Purchase prescription medications in advance if possible.

○□ Decide on non-prescription drugs that you may need.

○□ Purchase non-prescription drugs.

○□ List any related supplies or equipment that you might need.

○□ Purchase related supplies and equipment.

○□ Check any persistent ailments with your physician.

○□ Get paper backup of prescription requirements.

○□ Ensure you have cash if you will need to fill a prescription in early 2000.

○□ List all devices you use at home or at a medical facility (for example, dialysis machine, pacemaker, glucose testing equipment, inhalers, respirators, and so on).

○□ Investigate alternative power supply for devices in case of a power failure.

○□ Check with the manufacturers of the devices you rely on for a statement of Y2K compliance.

○□ Find out from your doctor what you can do if a necessary device doesn't work properly or fails.

○□ Make sure you have spare parts available for any devices.

○□ Learn how to make basic repairs for essential devices.

○□ Discuss these issues with members of your household.

○□ Build a basic first-aid kit and train your family in its use.

SANITATION

○□ Decide on your 'sanitation strategy.'

○□ Purchase portable toilet.

○□ Purchase appropriate supplies for portable toilet.

○□ Make sure you have soap and bleach.

COMMUNICATIONS

○□ Determine your alternative communications strategy.

○□ Check with neighbours to discover the nearest cellular phone.

○□ Make sure you have a battery-powered radio.

○□ Purchase sufficient batteries.

TRANSPORTATION

○□ Plan to stay where you party if necessary.

○□ Plan alternate routes home from New Year's Eve party.

○□ Arrange for someone reliable to check on your home while you are away.

○□ Ensure that you have a complete emergency kit in your car.

○□ Keep your car or truck fuel tank as close to full as possible.

FINANCES, RECORDS, INSURANCE

○□ Contact your home insurance agent for clarification of your policy.

Obtain a copy of:
- ○☐ your mortgage agreement
- ○☐ your latest bank statement
- ○☐ loan payment records
- ○☐ your credit card information
- ○☐ proof of rent payment
- ○☐ your insurance agreements
- ○☐ your birth and marriage certificates
- ○☐ divorce papers and agreements
- ○☐ your RRSP statements
- ○☐ social security information
- ○☐ your will
- ○☐ any contracts
- ○☐ other critical documents.
- ○☐ Secure your passports.
- ○☐ Secure your copy of your tax returns.
- ○☐ Decide how much cash you wish to have on hand.
- ○☐ Choose safe places to store your cash.
- ○☐ Withdraw cash in small amounts until you reach your goal.
- ○☐ Investigate longer-term financial strategies.

Y2K-SENSITIVE DEVICES

Check compliance of:
- ○☐ your programmable thermostat
- ○☐ your security system
- ○☐ your electronic locks
- ○☐ your sprinkler systems
- ○☐ your PC hardware and software.

TOWN AND COUNTRY

High rises and apartments
- ○☐ Arrange a tenants' Y2K meeting.
- ○☐ Check with landlord on Y2K plans for the building.
- ○☐ Initiate a Y2K action group.

Neighbourhood and community
- ○☐ Talk to your friends and neighbours about Y2K preparations.
- ○☐ Contact your local emergency preparedness office for local information.

SUPPLIES AND TOOLS

Purchase adequate supplies of:
- ○☐ toilet paper
- ○☐ tissues
- ○☐ soap
- ○☐ feminine hygiene supplies
- ○☐ baby supplies such as diapers and wet wipes
- ○☐ birth control supplies
- ○☐ vision-care products
- ○☐ dental-care products
- ○☐ plastic bags of various sizes
- ○☐ household disinfectant
- ○☐ pet litter.

Make sure you have:
- ○☐ a sewing kit
- ○☐ batteries
- ○☐ a utility knife
- ○☐ a shut-off wrench for water and gas
- ○☐ pliers
- ○☐ duct tape.

EMERGENCY ACTION PLAN

WHEN TO CONSIDER GOING TO A SHELTER

As a rule people don't like moving into emergency shelters; they prefer to stick it out at home. Even Ice Storm '98 drove only a small percentage of people to shelters, and some of those were moved under duress. But there may come a time when moving to an organized emergency evacuation centre is the prudent thing to do. Know the limits of your plans and try to recognize when you begin to reach them. Some indications might be:

* You find that you cannot cope with the situation.
* Your drinking water supply is getting dangerously low.
* Your food supplies are getting low and there are still no shops open.
* Your house or survival space is not adequate for your needs.
* You wake up cold and shivering in the morning and have no extra covers.
* Any member of your household begins showing signs of hypothermia.
* Any member of your household begins showing signs of CO poisoning, or your CO detector sounds its alarm.
  Remember: get the sufferer into the fresh air immediately, then seek medical attention. Do not re-enter your home without professional assistance, and if you cannot absolutely guarantee its safety, you must move somewhere else.
* You're feeling unmanageably stressed, lonely, or isolated— in other words, you have 'cabin fever.'
* The area that you are living in is evacuated by the police or military.

IF YOU DO HAVE TO LEAVE YOUR HOUSE

* If you have not already done so, and if you have time, cold-proof your dwelling before you leave, or arrange for someone to take care of it soon after you leave.
* Shut off the electricity and drain the plumbing if you haven't already done so.
* Pack and take adequate clothing and bedding, and what food and water you can carry.
* Pack some reading material, and games for the kids.
* Take your first-aid kit.
* Leave a note explaining where you are.
* Lock your house as you leave.

POWER LOSS

Stay calm. See if every house or building is affected.

If you haven't already read the detailed instructions, go to page 16.

SAFETY FIRST WITH ELECTRICITY

* Turn off the main breaker.
* Unplug sensitive electric appliances.
* Listen to your battery-powered radio for bulletins.

STAY WARM

If you have a backup heating unit, get it going before the house gets too cold.

DRAIN THE PLUMBING

* When the temperature in your house drops below 4°C, it is time to drain your water pipes.
* Turn off the water main where it enters the house.
* Starting at the top of the house, drain the water from your plumbing system.
* Drain water in dishwasher and other appliances.
* Add RV antifreeze to all traps, toilet bowls, sinks, drains, and faucets.